AF213970

An Introduction to Food Virology

An Introduction to Food Virology

Nigel Cook
Jorvik Food and Environmental Virology Ltd.
York, UK

David Rodríguez-Lázaro
Department of Biotechnology and Food Science
University of Burgos, Spain

Artur Rzeżutka
Department of Food and Environmental Virology
National Veterinary Research Institute
Puławy, Poland

Martin D'Agostino
Director, ViroSafety LTD
Stratford-upon-Avon, UK

WILEY

This edition first published 2026

Registered Offices
John Wiley & Sons, Inc., 111 River Street, Hoboken, NJ 07030, USA
John Wiley & Sons Ltd, New Era House, 8 Oldlands Way, Bognor Regis, West Sussex, PO22 9NQ, UK

For details of our global editorial offices, customer services, and more information about Wiley products visit us at www.wiley.com.

The manufacturer's authorized representative according to the EU General Product Safety Regulation is Wiley-VCH GmbH, Boschstr. 12, 69469 Weinheim, Germany,
e-mail: Product_Safety@wiley.com.

Wiley also publishes its books in a variety of electronic formats and by print-on-demand. Some content that appears in standard print versions of this book may not be available in other formats.

Library of Congress Cataloging-in-Publication Data:
Names: Cook, Nigel author | Rodríguez-Lázaro, David author | Rzeżutka, Artur author | D'Agostino, Martin author
Title: Introduction to food virology / Nigel Cook, David Rodriguez-Lazaro, Artur Rzeżutka, Martin D'Agostino.
Description: Hoboken, NJ: Wiley, 2026. | Includes index.
Identifiers: LCCN 2026024264 (print) | LCCN 2026024265 (ebook) | ISBN 9781394292899 paperback | ISBN 9781394292912 adobe pdf | ISBN 9781394292905 epub
Subjects: LCSH: Food–Microbiology | Virus diseases
Classification: LCC QR115 .C66 2026 (print) | LCC QR115 (ebook)
LC record available at https://lccn.loc.gov/2026024264
LC ebook record available at https://lccn.loc.gov/2026024265

Cover Design: Wiley
Cover Image: Soleskz/Shutterstock, KATERYNA KON/SCIENCEPHOTO LIBRARY/Getty Images, Chris Ryan/Getty Images, ConnectImages/Getty Images, kikisora/stock.adobe.com, Andrei Starostin/stock.adobe.com, Thanuch/stock.adobe.com, Ilie/stock.adobe.com

Set in 9.5/12.5pt STIXTwoText by Straive, Pondicherry, India

Contents

Foreword

By Professor Don Schaffner – Chair of Food Science, Rutgers University, USA

When Nigel Cook asked me to write the foreword for this book, I immediately said yes. Then I discovered that I didn't even know what a foreword was supposed to be, or how to spell it correctly (I'll come back to these points later). I think this is a fitting way for me to start this foreword, as it's also the same way that I became a 'sort of' virologist. We've been doing microbial risk assessment in my lab for quite some time, and eventually that included risk assessments for foodborne viruses. It wasn't until I had a graduate student that wanted to do something approaching real virology that I really started to learn more virology. When my student Robyn Miranda came to me and explained that we could easily start doing virology with a bacteriophage MS2 as a surrogate for norovirus, I said 'OK, as long as you can figure out how to do that', which of course she did. Then we had a pandemic, and people were very concerned about SARS-CoV-2, and whether it could be spread by food, so we took that as an opportunity to begin work with another bacteriophage called Phi 6. That line of research led to rekindling a collaboration with a former undergraduate student in my lab, Siobhan Duffy, who is now a full-fledged 'real' evolutionary virologist.

The papers that we've published on viral risk assessment have led to work with the World Health Organization and the Food and Agriculture Organization of the United Nations, and the chance to rub shoulders with more smart 'real virologists'. Those meetings and conversations have been an amazing experience. It was through those collaborations that I met Nigel, as well as many other dedicated, hard-working real virologists. The work that these virologists do is incredibly impressive and, by comparison, makes the work that I do in quantitative microbial risk assessment (making a few equations and running some simulations) look simple. It has been a steep learning curve for me to become fluent in the world of virology, but I have accomplished that by talking with smart virologists and reading the literature. Now it's even easier, and if I was going to begin my journey today, I would start by reading this book!

Nigel, Martin, David, and Artur have done a wonderful job writing this Introduction to Food Virology. As they say in the preface (which is not the same as a foreword), they wrote this book for the non-specialist who wishes to gain a basic knowledge of this growing branch of microbiology. I will add this is without a doubt a very fast-growing branch of microbiology. When I first learned about viral foodborne disease, it was a bit of a curiosity, featuring these weird little bundles of chemicals that were not alive but could hijack actual living cells to reproduce themselves in the world. Now even the young food microbiologist just starting out learns that the overwhelming number of cases of foodborne disease in many countries around the world is likely caused by norovirus. While most cases of norovirus are not fatal, the public health costs are tremendous. But, of course, not all viruses are the same, and while norovirus commonly causes mild illness, the same cannot be said for hepatitis A virus. And while there are almost a dozen viruses discussed in this book, I recently started learning about hepatitis E virus, and my mind was blown. This obscure hepatitis E virus may cause many more illnesses than we even know about until we start looking for them.

The purpose of a foreword is to introduce the authors and work to readers, and to tell readers why they should read the book, so I hope that I've done that. And in case you are ever asked to write a foreword, just remember it is a 'fore' (like they shout in golf) 'word' or words that come before the actual words of the authors. What are you waiting for? Get reading.

About the Authors

Dr. Nigel Cook

Dr. Nigel Cook studies the transmission of pathogens, particularly enteric viruses, through foods and the environment, and the development and application of nucleic acid amplification-based detection methods. He is a graduate of the University of Dundee. After postdoctoral research in the Universities of Aberdeen and Leicester, he moved to the Central Science Laboratory (CSL) at the Food Science Laboratory, Torry, Aberdeen, in September 1994, before relocating to new facilities in York in 1997. He led a research laboratory in CSL and its successor institutes until July 2017 and is now an independent researcher. He is a Fellow of the Institute of Science and Technology and a Board Member of the International Society for Food and Environmental Virology. He is currently involved in various international standardisation activities and leads a working group on collaborative initiatives within a European COST Action network on zoonotic hepeviruses. He holds a Guest Professorship at KU Leuven, Belgium.

Prof. David Rodríguez-Lázaro, DVM, MSc, PhD

Professor David Rodríguez-Lázaro is a Full Professor of Microbiology, Head of the Microbiology Division and Director of the Research Centre for Emerging Pathogens and Public Health at the University of Burgos (Spain), where he also leads an Excellence Research Unit in "One Health" Microbiology. He possesses more than 25 years of experience in the establishment of reliable strategies for detection and control of important food-borne pathogens from environmental sources and various types of foodstuffs.

He has participated in a large number of international research projects and has published more than 200 international scientific papers, books and book chapters regarding to Food Microbiological Safety. He is member of the editorial board of the most relevant journals on the Food and Applied Microbiology and is the editor-in-chief of the Journal *Food Analytical Methods* (Springer).

Prof. Artur Rzeżutka, DVM, PhD, ScD

Professor Artur Rzeżutka is a senior scientist, virologist at the Department of Microbiology of Food and Feed of the National Veterinary Research Institute in Puławy, Poland. His research activities encompass a method development and their application for virus detection in food chain, studies on zoonotic viruses with a special interest in virus epidemiology in farm animals and wildlife. In the Institute, he was heading the Department of Food and Environmental Virology and established the National Reference Laboratory for foodborne viruses in food of animal origin. As an expert, he participated in several scientific committees and professional bodies (Codex Alimentarius Commission, ISO/CEN, Med-Vet-Net Association, COST FA Committee). He was involved in a number of national and international research projects on viruses in animals, food and the environment. He acts as an editorial board member for five national and international scientific journals. Prof. Rzeżutka is a member of Polish Society of Veterinary Science, the European One Health Association and EFSA working group.

Dr. Martin D'Agostino, PhD, FRSB, FIFST, MIScT

Dr. Martin D'Agostino is a highly experienced Food Virologist with over 30 years of expertise in laboratory-based research, specialising in the transmission, detection and prevention of foodborne viruses within food supply chains. As Director of ViroSafety Ltd, Martin provides bespoke training, crisis management and consultancy services, working closely with clients across the food industry, government agencies and academic institutions.

Previously, Martin served as Principal Virologist at Campden BRI, where he established the UK's first UKAS-accredited laboratory for the detection of norovirus and hepatitis A virus in fresh and frozen produce. He has led significant national and international research projects, contributes extensively to scientific publications and plays a key role as an expert advisor on ISO/CEN committees. Martin is a member of the Royal Society of Biology and a Fellow of the Institute of Food Science and Technology.

Preface

This introduction to food virology is written for the non-specialist who wishes to gain a basic knowledge of this growing branch of microbiology. Without going into excessive technical detail, it aims to present a clear overview of the main principles and features of the field of food virology. Each foodborne virus (the ones of main concern and those less commonly identified) is outlined, with simple descriptions of its structure, how it is transmitted, and what disease symptoms it can cause. Sections on how foods can become contaminated, and on virus survival and persistence, explain how these microbial entities have an impact on the food supply chain. Descriptions of outbreaks of foodborne viral disease are presented as informative examples. A chapter on how they may be eliminated or prevented from entering a food chain highlights some of the challenges faced when trying to control the transmission of foodborne viruses. Finally, a discussion of challenges which will be faced in the future, including the potential emergence of new viral threats, illustrates the ongoing importance of food virology.

This book will be useful to food safety managers, regulators, food producers, food service personnel, environmental health officers and inspectors, laboratory staff new to analytical virological methods, students, academic staff teaching aspects of food safety, and others who are involved in or interested in the microbial safety of the food supply chain and its impact on public health.

Where data (e.g. annual number of disease cases) are mentioned within this book, they are current at the time of publishing. For ease of reading, the sources (scientific publications, etc.) are not cited or referenced within the text, but are supplied at the end of each chapter. As far as possible, we have used sources which are in the public domain, e.g. through open-access; however, some of the source publications are available only through journal subscriptions. In many cases, we have cited comprehensive review articles in which the interested reader may find the specific references for the mentioned data. Additionally, recommended examples of further reading are supplied at the end of each chapter. These are either freely available online, e.g. open-access publications, or are easily obtainable through a library. They have been chosen to point the reader towards information which will be regularly updated (e.g. from the World Health Organization [WHO], Centers for Disease Control and Prevention [CDC], etc.), or scientific publications (e.g. reviews) which contain comprehensive general information while allowing identification of specific sources of data.

Glossary

Amplicon The product of a polymerase chain reaction. It is a DNA molecule which contains a nucleotide sequence which mirrors that of the target microorganism's nucleic acid target sequence.

Antigenicity The capacity to stimulate the production of antibodies or the capacity to react with an antibody.

Calicivirus A virus belonging to the family *Caliciviridae* (e.g. norovirus). It is a non-enveloped, positive-stranded RNA virus with an icosahedral capsid.

Capsid The protein shell of a virus. The capsid provides structural support and protection to the viral genome and is essential for the virus's ability to infect host cells.

Ct Cycle to threshold; the term that indicates the point in a real-time PCR reaction where the increase in amplicons becomes exponential.

Deoxyribonucleic acid (DNA) The molecule that carries the genetic instructions for the development, functioning, growth, and reproduction of all known organisms and many viruses.

Elimination Removal of infectious virus by a deliberately applied procedure.

Endemicity The prevalence of a disease or disease agent in a specific area or group of people.

Gastroenteritis Gut inflammatory diseases that are associated with diarrhoea and/or vomiting.

Genome The complete set of genetic material, including all the genes and non-coding sequences, contained within an organism's DNA or RNA. It serves as the blueprint for the development, functioning, and inheritance of living organisms.

Hepatitis An inflammatory disease of the liver.

Immunoglobulins Antibodies produced by the immune system to identify and neutralise foreign substances, including viruses.

Inactivation The decline in virus infectivity effected by a deliberately applied procedure.

Kb kilobase, one thousand nucleotides (the building blocks of nucleic acid) along a strand.

kGy Kilo Gray, a measure of radioactivity.

Mastermix The combined components of a polymerase chain reaction (PCR) prior to the addition of the target.

MegaPascal (MPa) A measure of force per unit area, defined as 1,000,000 newtons per square metre.

mRNA A single-stranded ribonucleic acid present in the infected cell which carries information for synthesis of viral proteins.

Nested PCR (nPCR) A biochemical procedure in which the amplified DNA sequence serves as target sequence in subsequent amplification.

Open reading frame (ORF) The part of a gene that is directly translated to a protein.

Persistence The retention of infectivity of virus particles in the environment.

Picornavirus A virus belonging to the family *Picornaviridae* (e.g. hepatitis A virus). It is a non-enveloped, positive-stranded RNA virus with an icosahedral capsid.

Polymerase Chain Reaction (PCR) A biochemical procedure which results in massive amplification of a target DNA sequence.

Positive-sense RNA virus A virus whose genetic information consists on a single strand of RNA that is the positive (or sense) strand which encodes mRNA and protein.

Quality-Adjusted Life Years (QALY) A measure used in health economics to assess the value of medical interventions. One QALY equates to one year in perfect health. This measure considers both the quantity and quality of life generated by healthcare interventions, enabling comparison of the value of different health outcomes.

Quantitative Microbiological Risk Assessment (QMRA) A systematic approach to quantify the risk of foodborne infectious disease.

Reverse Transcription (RT) A biochemical reaction in which a DNA mirror copy of an RNA sequence is formed.

RT-PCR A reverse transcription polymerase chain reaction. A biochemical procedure where target RNA sequences are copied to complementary DNA sequences prior to their amplification by PCR.

RTqPCR An RT-PCR in which the PCR component can allow quantitation of the transcribed DNA.

Serotype A distinct group within a species of virus, classified based on its common possession of surface structures with antigenic properties.

Survival The persistence of virus infectivity when no elimination treatment is deliberately applied.

Virion A complete virus particle (consisting of an RNA or DNA core with a protein coat, and sometimes with a lipid envelope). This is the infective form of a virus.

Viremia Virus presence in the blood of the infected host.

Zoonosis A disease transmitted from animals to humans. The disease agent may not necessarily cause disease in the animal host.

1

Introduction

Foodborne diseases are one of the most serious problems in public health. Awareness of this issue has increased in the past decades with several prominent outbreaks and better recognition of food-associated disease cases. Globalisation of the markets and the resulting increase of international food transportation has also escalated the potential for the spread of foodborne pathogens. It has been estimated that about 4.5 billion cases of gastroenteritis occur worldwide annually, causing about 4 billion litres of diarrhoeal discharges, roughly equivalent to all the water that passes over Victoria Falls in one hour. It has been estimated that in the United States, the most prominent foodborne pathogens (including *Salmonella* spp. and protozoan parasites such as *Toxoplasma gondii*) annually cause 9.9 million cases, resulting in 53,000 hospitalisations and over 900 deaths. Noroviruses (see Chapter 3) are responsible for the greatest number of cases (5.5 million) and the highest number of hospitalisations (22,400). The cost to the US economy, associated with medical expenses and lost working hours, runs into billions of dollars. Analogous figures have been estimated for other countries such as the United Kingdom and Australia. Therefore, the impact of foodborne pathogenic microorganisms on public health systems and national economies is very considerable. However, the importance of viral foodborne diseases worldwide remains largely underestimated, due to difficulties in accumulating accurate data on the incidence of outbreaks and cases. Isolated cases often go unnoticed, especially when the affected individual does not report to a medical practitioner. Nonetheless, the reported incidence of cases of enteric viral diseases associated with food consumption has increased in recent years in developed countries, and enteric viruses have become the main cause of foodborne illness, causing the majority of episodes.

Viruses, the Microbial Enigma

> *'A virus is nothing more than a piece of nucleic acid surrounded by bad news.'*
>
> – Peter Medawar.

The term 'virus' comes from a Latin word which refers to venom or poison and likely originally derives from the Latin *viscum*, meaning a sticky substance or ligand. It was used for the first time in the English language in 1392 and associated with infectious diseases in 1728 in reference to a venereal disease.

Viruses are parasitic entities. They are incapable of replication without a host and are therefore classed as obligate intracellular parasites. They are very simple in structure, basically comprising genomic material surrounded by a protein shell (the capsid) and sometimes with an outer lipid envelope (Figure 1.1).

The viruses which can infect humans and animals are extremely small, ranging in size from 17 nm (porcine circovirus) to 300 nm (poxvirus). The largest viruses are one third the size of the smallest bacterium. None but the very largest viruses can be seen using conventional light microscopes and

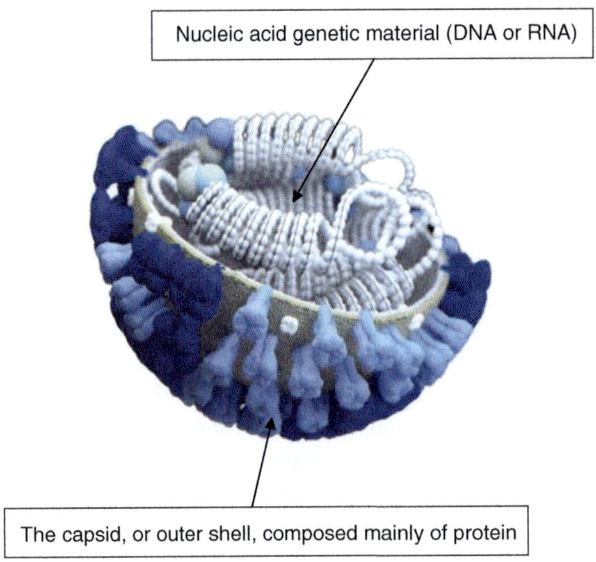

Figure 1.1 The basic structure of a virus.

are visible only by using an electron microscope. Nonetheless, these tiny organisms have an immense impact on society. Each single living species (including animals, plants, and bacteria) has at least one virus type which can infect it.

Viruses are classified into various divisions (e.g. genus, family, type) based on their characteristic structures. The main division is based on the nature of their genomic material, i.e. DNA or RNA; the secondary division is based on whether the virus particle has an envelope surrounding the capsid. Viruses, although classed as microorganisms, are not alive in the strictest sense: they have no intrinsic metabolism. They require the metabolic apparatus of a host organism to complete their cycle of existence and produce more copies of themselves. This hijacking of the host's internal systems can in many cases have an adverse effect on the host, an effect which becomes manifest as disease symptoms. Viruses that cause disease can be transmitted from host to host by several routes, including contact with host body fluids, excretions, and exhalations. Some viruses, e.g. dengue virus, can be transmitted by insects, whereas others such as rabies virus are transmitted via infected mammals. Other routes of transmission include air (e.g. for influenza virus) and water (e.g. for astrovirus). Viruses which are predominantly shed in the faeces of infected persons and are then taken in by the next person through the mouth into the gastrointestinal tract are said to have a faecal–oral route of transmission. All faecal–orally transmitted viruses have the potential to be foodborne.

There are several types of viruses which are capable of transmission through contaminated foods (Table 1.1). Foodborne viruses include agents of gastroenteritis, as well as agents of more serious diseases such as hepatitis and meningitis. Most of these viruses belong to the families *Adenoviridae*, *Caliciviridae*, *Hepeviridae*, *Picornaviridae*, and *Reoviridae* and are termed 'enteric viruses' since they replicate in the internal organs of infected animals or humans (Figure 1.2) and are excreted from their gastrointestinal tracts, predominantly via faeces, although vomit can also play a role.

The excreted viruses are then spread through the environment by various routes, including water which has been contaminated by sewage and/or agricultural waste. Contact between, for example, growing food crops and contaminated water can result in viruses contaminating the food supply chain. Direct contact with virus-infected persons is another major means by which viral contamination of food can occur. All these aspects will be described in detail.

Table 1.1 The main foodborne viruses infecting humans.

Virus	Family	Approximate size (nm)	Genome	Host	Major symptoms
Adenovirus	*Adenoviridae*	100	dsDNA	Human	Diarrhoea, vomiting
Astrovirus	*Astroviridae*	28	ssRNA	Human	Diarrhoea, vomiting
Coxsackie virus	*Picornaviridae*	28	ssRNA	Human	Meningitis, muscle ache, paralysis (rare)
Enterovirus	*Picornaviridae*	28	ssRNA	Human	Meningitis, muscle ache, paralysis (rare)
Hepatitis A virus	*Picornaviridae*	27–32	ssRNA	Human	Inflammation of the liver
Hepatitis E virus	*Hepeviridae*	27–34	ssRNA	Human, pig, deer, other animals	Inflammation of the liver
Norovirus	*Caliciviridae*	34	ssRNA	Human	Projectile vomiting, diarrhoea, malaise
Poliovirus	*Picornaviridae*	28	ssRNA	Human	Paralysis, meningitis, muscle ache
Rotavirus	*Reoviridae*	75	dsRNA	Human, animals	Diarrhoea, vomiting
Sapovirus	*Caliciviridae*	34	ssRNA	Human	Vomiting, diarrhoea
Tick-borne encephalitis virus	*Flaviviridae*	50	ssRNA	Ticks, via dairy animals	Neurological

Key: ds, double-stranded; ss, single-stranded.

History of Food Virology

Humanity has been suffering from viral diseases since before the dawn of civilisation, but only within the past 150 years have technological advances led to the discovery of these agents and their characterisation. The discovery of viruses was performed independently in two laboratories in the

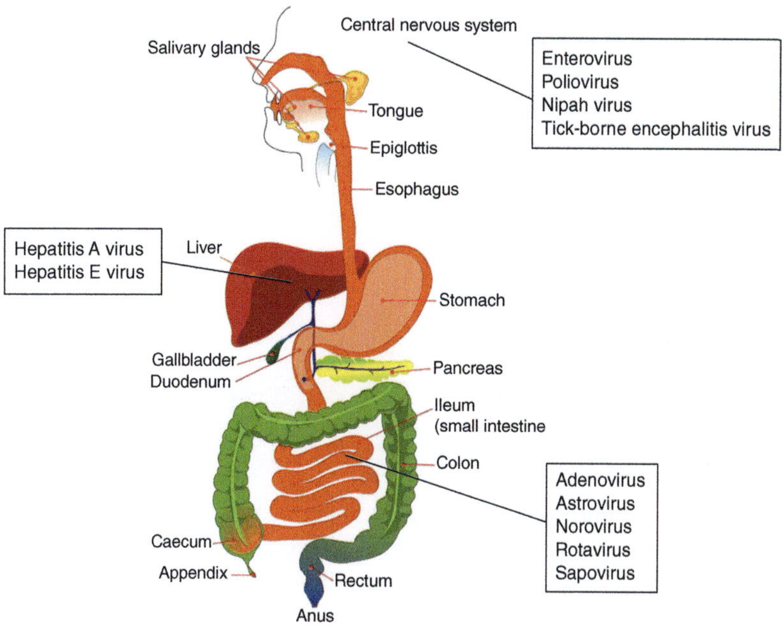

Figure 1.2 The major sites of replication of enteric viruses. *Source:* Mariana Ruiz LadyofHats / Wikimedia Commons / Public Domain and has been modified to include the virus labels.

Figure 1.3 Dmitri Iosifovich Ivanovsky, known as the 'Father of Virology'. *Source:* Wikimedia Commons / CC BY 4.0.

late nineteenth century by researchers Dmitri Ivanovski (1864–1920) in Russia (Figure 1.3) and Martinus Beijerinck (1851–1931) in the Netherlands (Figure 1.4). In both cases, they used a similar strategy: extracts of infected samples were passed through filters with pore diameters smaller than those

Figure 1.4 Martinus Willem Beijerinck. *Source:* Courtesy of Wellcome Collection.

of bacterium. Dimitri Ivanovski, in 1892, studying tobacco plant mosaic disease, found that the extract of leaves of infected tobacco plants remained infectious after passing it through a filter, although initially the effect on the plants was thought to be caused by a bacterial toxin. In 1898, Martinus Beijerinck, unaware of the previous studies by Ivanovski, obtained similar results, confirming that the active principle was a new form of infectious agent. He called these agents *contagium vivum fluidum* (live contaminating fluid). That same year, Friedrich Loeffler and Paul Frosch, both collaborators of Robert Koch in Germany, obtained similar results for an animal virus, the foot-and-mouth disease virus. These investigators ruled out the presence of a bacterial toxin on the basis that foot-and-mouth disease could be passed from one animal to another and concluded that the agent was able to replicate; moreover, it was smaller than the smallest known bacterium and, therefore, beyond the resolution power of the microscopes then available.

Hippocrates described an illness characterised by episodes of jaundice (that may have been viral hepatitis), and the disease was recognised as contagious in the Middle Ages, particularly during wars. During the wars of the early and mid-twentieth century, outbreaks often vast in scale occurred in the armies of several nations. During the Second World War and subsequently, human volunteer studies established that viruses were the cause of epidemic hepatitis. In due course, hepatitis A became recognised as an infection which was acquired through ingestion of viruses, in contrast to the blood-borne hepatitis B. Hepatitis A virus (HAV) was first identified in 1973 by Dr. Stephen Feinstone and his colleagues who isolated the agent from the stools of infected patients.

The first recorded outbreak of foodborne disease attributable to a virus occurred in the early twentieth century. In 1914, four cases of poliomyelitis (a paralytic disease) occurred in an English community among children who drank milk from a common source, although how the milk had become contaminated was not determined. Between 1914 and 1949, nine additional outbreaks of foodborne poliomyelitis were reported in the United Kingdom and the United States. The recognition of poliovirus infection in these outbreaks was based on clinical symptoms, as effective laboratory diagnostic methods were not yet available. Raw milk was the apparent vehicle in five of these additional outbreaks, and pasteurised milk that might have been contaminated after warming was implicated in two more. Raw milk was also the recorded vehicle of a hepatitis outbreak comprising 14 cases in Scotland in 1943. What may have been the first outbreak of enteric viral disease associated with consumption of shellfish occurred among people who ate raw oysters in Sweden between 1955 and 1956 and caused more than 600 cases of hepatitis A. In 1972, the first evidence of a viral cause for human diarrhoeal disease was obtained. An outbreak of gastroenteritis had occurred at an elementary school in Norwalk, Ohio, in 1968. The source of the outbreak was unknown at the time, but fortuitously, samples of the infected individuals' stools were kept frozen. Four years later, Albert Kapikian and collaborators analysed these samples by electron microscopic examination and discovered Norwalk virus, the prototype of the genus *Norovirus*.

Despite the now-public awareness of the impact of virus contamination of the food supply chain, there was little interest in foodborne viruses until the early years of the millennium, with the scope of food microbiological safety predominantly focussed on bacteria and parasites. But since then, the increasing reports of foodborne viral outbreaks and the availability of more sensitive detection technologies have generated extensive research, conducted both nationally and internationally, which has resulted in a growing body of knowledge within the discipline of food virology. The leading pioneer of this discipline was Professor Dean O. Cliver in the United States, coordinator of the first research program on the transmission of viruses through food (Figure 1.5).

A milestone was the creation of an international scientific society dedicated to the study and promotion of this discipline: the International Society for Food and Environmental Virology (ISFEV; www.isfev.org). The first president of ISFEV is Professor Albert Bosch (Figure 1.6) of the University of Barcelona. The Society has an international scientific journal dedicated to this discipline: *Food and Environmental Virology*, edited by Springer-Nature, and holds an international conference every two years.

Figure 1.5 Professor Dean O. Cliver. *Source:* Photo courtesy of Gary Richards.

Figure 1.6 Professor Albert Bosch. *Source:* Courtesy of A. Bosch.

Transmission of viruses via food is now recognised as a major public health concern and also as a significant challenge to the food industry. The following chapters will provide information on the most important foodborne virus types, how they can contaminate foodstuffs, what techniques can be used to detect contamination along the food supply chain, and what measures can be taken to control virus contamination. The reader will gain an understanding of the various challenges posed by foodborne viruses and be given an insight as to how these challenges may be met.

Sources of Quoted Data

Norovirus infectious dose: Atmar, R.L. et al. (2014). Determination of the 50% human infectious dose for Norwalk virus. *The Journal of Infectious Diseases* 209: 1016–1022. Teunis, P.F. et al. (2008). Norwalk virus: how infectious is it? *Journal of Medical Virology* 80: 1468–1476.

Total gastroenteritis cases: GBD 2016 Diarrhoeal Disease Collaborators (2018). Estimates of the global, regional, and national morbidity, mortality, and aetiologies of diarrhoea in 195 countries: a systematic analysis for the Global Burden of Disease Study 2016. *The Lancet Infectious Diseases* 18: 1211–1228.

USA foodborne pathogens annual incidence: Walter, E.J.S. et al. (2025). Foodborne illness acquired in the United States—major pathogens, 2019. *Emerging Infectious Diseases* 31: 669–677.

Enteric viruses as main cause of foodborne illness: Sair, A.I. et al. (2024). Human enteric viruses as causes of foodborne disease. *Comprehensive Reviews in Food Science and Food Safety* 1: 73–89.

Further Reading

Carter, J. (2013). *Virology: Principles and Applications*, 2e. Wiley.

Flint, S.J. et al. (2020). *Principles of Virology*, 5e. ASM Press.

Knipe, D.M. and Howley, P.M. (2013). *Fields Virology*, 6e. Lippincott Williams & Wilkins.

Koopmans, M.P.G. et al. (2008). *Food-borne Viruses: Progress and Challenges.* ASM Press.

Radetsky, P. (1995). *The Invisible Invaders: Viruses and the Scientists Who Pursue Them.* Little Brown and Company.

UK Health Security Agency (2025). Foodborne outbreaks of norovirus in England, 2015 to 2023. *Health Protection Report* 19 (9).

2

The Foodborne Viruses

Any infectious virus shed in the faeces or vomit of an infected person or animal, particularly those transmitted via the faecal–oral route, has the potential to be foodborne. Foodborne viruses have a set of general features in common, and these features are shared by several virus types (see Box 2.1). The most significant foodborne viruses, in terms of number of reported disease cases/severity of infection, are norovirus, hepatitis A virus (HAV), and hepatitis E virus (HEV) (Table 2.1).

Norovirus is significant due to the high number of disease cases it is estimated to cause annually. The number of reported cases has increased in recent years, in part due to the availability of effective diagnostic procedures, and in part possibly through increased consumption of the types of foods most at risk of contamination, e.g. leafy green vegetables (see Chapter 3). HAV is currently recognised as the most important cause of infectious hepatitis. In many countries, the level of immunity to this virus is falling in the population due to increased public hygiene standards, and this can create an opportunity for large foodborne outbreaks to occur. HEV is becoming increasingly recognised as a zoonotic disease agent, as it is carried by domestic pigs and some wildlife species. The number of reported cases of apparent zoonotic infection with HEV is rising in several countries. Other viruses which have occasionally been reported to cause foodborne disease include rotavirus, sapovirus, and tick-borne encephalitis virus, and there are emerging virus species which may also possess the potential for foodborne transmission, such as Nipah virus (Table 2.2).

The potential of this cavalcade of infectious agents to contaminate the food supply poses a serious threat to the food industry and the public health sector. The key challenges are driven by limitations in current detection methods, gaps in surveillance, and in implementing effective control measures across complex globalised food supply chains. Overcoming these

Box 2.1 General Characteristics of Foodborne Viruses

Do not possess a lipid envelope
Faecal–oral transmission
Do not multiply (replicate) in or on food
Very low infectious dose (from 10 to 1000 virus particles)
Can be shed in very high numbers (e.g. 10^8 per gram or ml)
Very stable outside the host
Resistant to acidic and basic pH and several disinfectants

challenges will require an increased awareness of the threat and a coordinated use of resources to drive research, surveillance, and prevention.

Norovirus

Norovirus, formerly known as small round structured virus (SRSV) and Norwalk-like virus (NLV), belongs to the virus family *Caliciviridae*. Caliciviruses were named after the superficial cup-shaped depressions (calix = a cup) observed via the electron microscope. This family is composed of five genera: *Norovirus*, *Sapovirus*, *Lagovirus*, *Vesivirus*, and *Nebovirus*. Only *Norovirus* and *Sapovirus* contain human pathogens; the other three infect animals. Noroviruses are non-enveloped icosahedral viruses with a diameter of ~34 nm (Figure 2.1).

The viral genome is a ~7.5-kb positive-sense RNA that contains three open reading frames (ORFs; analogous to genes) which encode structural (capsid) and non-structural (enzymes) proteins (Figure 2.2).

The norovirus capsid is composed of 180 copies of the major capsid protein (VP1) organised in 90 dimers with icosahedral symmetry and up to 100 copies of the minor structural protein (VP2).

Noroviruses are genetically and antigenically very diverse. Based on differences in the VP1 capsid protein gene, they are currently classified into ten genogroups (Table 2.3).

The most common strains associated with gastrointestinal disease in humans belong to the GI and GII genogroups and, to a lesser extent, GIV. Genogroups are further subdivided into genotypes, which are genetic clusters based on the nucleotide sequence diversity of the area of ORF1 that encodes the VP2 protein.

Norovirus particles bind to receptors in the epithelial cells of the small intestine. These receptors are encoded by the human histo-blood group

Table 2.1 Features of the three most significant foodborne viruses: norovirus, HAV, and HEV.

Virus species	Infectious dose	Onset of symptoms	Characteristic symptoms of disease	Duration of symptoms	Long-lasting immunity conferred	Zoonotic potential
Norovirus	Approx. 18 viral particles	12–72 hours	Projectile vomiting, diarrhoea, fever, malaise	1–3 days	No	No
Hepatitis A virus	Unknown	15–50 days	Jaundice, fever, malaise, nausea, abdominal discomfort, dark urine	2–6 weeks (up to 6 months in some cases)	Yes	No
Hepatitis E virus	Unknown	4–5 weeks	Jaundice, fever, malaise, nausea, abdominal discomfort, dark urine	1–4 weeks (up to 6 months in some cases)	Yes	Yes

Table 2.2 Occasionally or potentially foodborne viruses.

Virus species	Main diseases produced	Foodborne outbreaks confirmed	Zoonotic potential
Adenovirus	Gastroenteritis, encephalitis, conjunctivitis, pneumonia	No	No
Astrovirus	Gastroenteritis	Yes	Possibly
Coronavirus	Gastroenteritis, pneumonia	No	Possibly
Enterovirus	Gastroenteritis, encephalitis, meningitis, cardiological disorders	Yes	No
Nipah virus	Encephalitis, pneumonia	Yes	Yes
Parvovirus	Gastroenteritis	No	No
Rotavirus	Gastroenteritis	Yes	Possibly
Sapovirus	Gastroenteritis	Yes	No
Tick-borne encephalitis virus	Neurological disorders	Yes	Yes

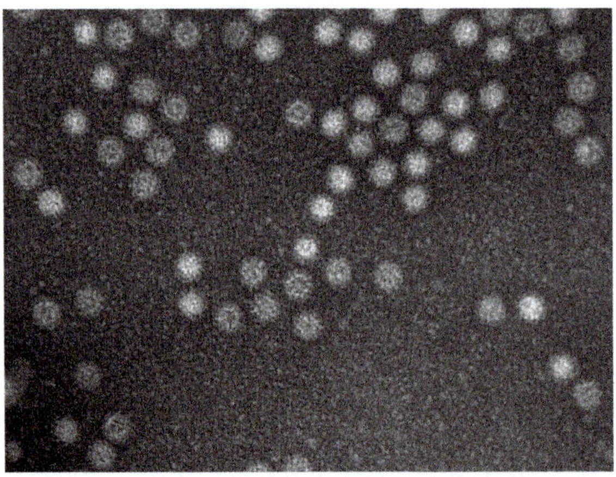

Figure 2.1 Norovirus particles, showing their icosahedral structure. *Source*: Katerina Kovač, University of Ljubljana.

antigen (HBGA) gene. Therefore, norovirus infection depends on the presence of specific HBGA receptors in the intestine in the susceptible host. The virus replicates within the intestinal cells, resulting in damage including blunting of villi (the finger-like structures located on cell surfaces which aid in the absorption of nutrients), cell death, and delays in

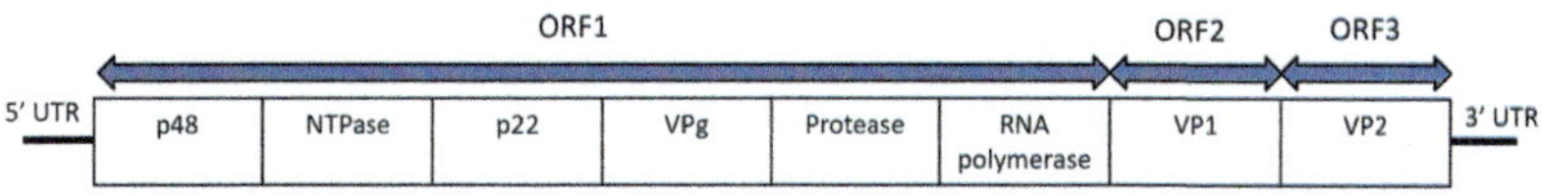

ORF: open reading frame
UTR: untranslated region
p48: function not yet known
NTPase: enzyme involved in virus replication
p22: promotes virus assembly
VPg: directs virus protein synthesis
Protease: enzyme involved in virus replication
RNA polymerase: enzyme involved in virus replication
VP1: major capsid protein
VP2: minor capsid protein

Figure 2.2 Structure of norovirus genome.

Table 2.3 Norovirus classification (genogroups recognized as of 2025).

Genogroup	Host species
I	humans
II	humans, pigs
III	bovines, mice
IV	pigs, humans
V	mice, macaques
VI	canines, felines
VII	canines
VIII	humans
IX	humans
X	bats

gastric emptying. The effects of this damage result in the characteristic clinical features of norovirus gastroenteritis.

Noroviruses infect all age groups, with particularly severe symptoms in the very young, the elderly, and the immunocompromised. The infectious dose (estimated using data from human volunteer studies) may be very low, possibly less than 20 infectious virus particles in susceptible individuals, although the difficulty in propagating and quantifying infectious norovirus (see Chapter 4) makes estimation challenging, and the actual infectious dose may be higher. The incubation period is 10–51 hours, and the symptoms are characterised by an acute onset of nausea, abdominal cramps, non-bloody diarrhoea, and myalgia. The most characteristic symptom of norovirus gastroenteritis, affecting around 50% of symptomatic individuals, is projectile vomiting, where vomitus is forcibly ejected from the sufferer and can travel several feet. Vomit from an infected person can contain on average 10^8 particles per ml, although 10^{10} particles per ml have been reported. The symptoms usually disappear within 2–3 days, but may last longer (e.g. 4–6 days) in immunocompromised persons and children younger than 11 years of age. It is estimated that approximately 16% of infected individuals never develop symptoms. Symptomatic and asymptomatic individuals can shed norovirus in their faeces for up to 3 weeks (in some individuals, e.g. those who are immunocompromised, it can be longer), and thus they can transmit the virus even after the disappearance of clinical symptoms. Shedding in faeces can also occur before the onset of symptoms (presymptomatic shedding). As with asymptomatic infection, pre- and post-symptomatic shedding of virus have considerable implications for food safety, as will be discussed in Chapter 3. On average, 10^8 norovirus particles are shed per gram of faeces from an infected person. Secondary transmission to other individuals can occur. Outbreaks caused by person-to-person transmission usually occur in closed or semi-enclosed environments, such as schools, hospitals, camps, nursing homes, resorts, prisons, cruises, or restaurants, where the virus, once introduced, spreads very rapidly. There is no specific treatment for norovirus gastroenteritis, although anti-vomiting and rehydration therapy is sometimes administered for severe cases. The disease is, in the great majority of cases, self-limiting in that the symptoms eventually resolve by themselves. However, severe disease, hospitalisations, and deaths, although uncommon, do occur. Post-infectious irritable bowel syndrome can occur in up to 13% of individuals. Infection with norovirus confers only short-term and strain-specific immunity, and a patient may be re-infected by another norovirus and develop symptoms, only a few months after the previous infection.

Hepatitis A Virus

Hepatitis A virus (HAV) is classified within the family *Picornaviridae*. Picornaviruses ('small RNA' viruses) were named from their possession of an RNA genome and a small capsid. The *Picornaviridae* family contains other genera, some of which, e.g. poliovirus, can be transmitted via food. The HAV particle has an icosahedral capsid of about 27–32 nm

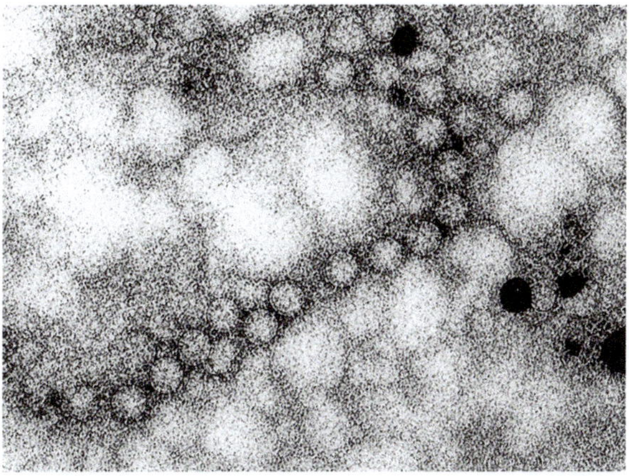

Figure 2.3 An electron micrograph of hepatitis A virus HM175, strain 43c.
Source: Albert Bosch and Rosa Pinto. Enteric Virus Laboratory, University of Barcelona.

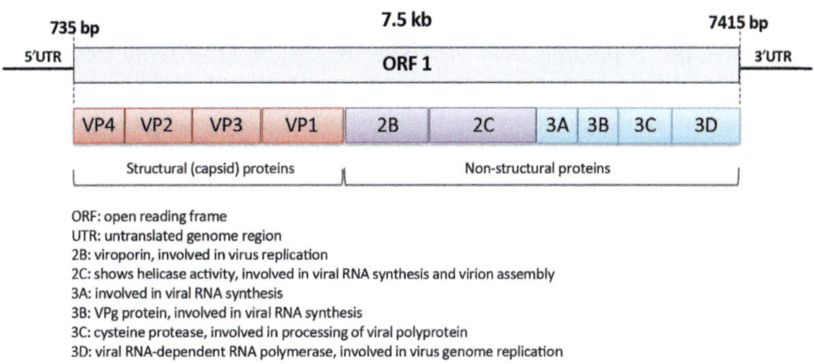

Figure 2.4 Structure of the HAV genome.

with no envelope (Figure 2.3) and contains a single-stranded RNA genome of approximately 7.5 kb.

The genome contains only one ORF; this encodes a single protein which is cleaved into 11 separate structural and non-structural proteins in the host cell. Each virus capsid is composed of 32 capsomers or protein subunits (Figure 2.4).

The classification of HAV strains is based on the nucleic acid sequences within the capsid protein genes. HAV can thereby be classified into six genotypes and subgenotypes. Genotypes I, II, and III have been isolated from humans, whereas genotypes IV, V, and VI are of simian origin. The virus genotypes of human origin are further divided into seven subgenotypes (IA, IB, IC, IIA, IIB, IIIA, and IIIB). The strains of subgenotype IA are the most prevalent worldwide. Co-circulation of subgenotypes IA and IB has been reported in some geographical areas.

As with other picornaviruses, recombination between HAV strains can occur. Recombination events can occur among the same genotype and between different genotypes. Also, minor sequence changes can occur through virus evolution (genetic drift). However, these changes do not result in the formation of new serological variants; thus, all currently known HAV strains have the same serotype. Consequently, in vaccinated individuals a single vaccine strain can induce the formation of protecting antibodies against HAV strains circulating on different continents. Furthermore, vaccination provides long-lasting immunity, and this could be advantageous for control of HAV contamination of food (see Chapter 6).

HAV transmission occurs via the faecal–oral route by direct contact with ill persons or asymptomatic young children and in rare cases by a sexual or parenteral route. Mainly due to the impracticality of conducting volunteer studies for this virus with its rather severe symptoms, the infectious dose is unknown, but is possibly very low. After ingestion, HAV replicates in the alimentary tract, and then the virus passes to the liver via the hepatic portal vein. As with norovirus, HAV has high resistance to the several natural barriers in the digestive tract it must overcome, such as the low pH of the digestive fluid present in the stomach and proteolytic enzymes and detergents present in the intestine. In the liver, viruses attach to receptors on the hepatocytes (liver cells) and then enter and replicate there. Virus can also spread between hepatocytes, by using the cellular protein transport system. After virus formation, the progeny viruses are secreted into the small vessels in the liver (canaliculi) which carry bile, from where they are transported via the bile duct to the intestines to be finally excreted in faeces. HAV presence in the liver cells triggers an inflammation, which is associated with an immune response leading to the organ's damage. The liver is the primary site of virus replication, but it can also occur in the epithelial cells of the bile duct, the small intestine, and the salivary glands.

The mean incubation period of the disease is around 30 days, with a maximum of 50 days. During active virus replication, infected persons remain generally asymptomatic. A viraemia precedes the development of symptoms, and viruses can be present in the blood of an infected person for several weeks before the onset of symptoms. Although HAV is considered as a non-enveloped virus type, recent studies show that it can circulate in the blood as quasi-enveloped particles (eHAV), coated by host cell membrane material. eHAV particles are formed during non-lytic release of the virus

Table 2.4 The main symptoms of hepatitis A.

Symptom	Description
Jaundice	Yellowing of the skin and eyeballs; the classic symptom of hepatitis
Fever	Elevated body temperature associated with viral infection
Malaise	A feeling of tiredness and general unwellness
Nausea	A sensation of discomfort in the stomach, often accompanied by an urge to vomit
Abdominal discomfort	Pain or discomfort in the upper right quadrant of the abdomen
Dark urine	Dark-coloured urine, often accompanied by pale stools

from infected cells, i.e. where the cells are not disrupted by the exiting virus particles. Despite observed morphological differences between these two virus forms, eHAV particles have similar infectivity as the non-enveloped particles. It has been shown that most virions present in the supernatant of infected cell cultures are also quasi-enveloped. For the symptoms of HAV infection, see Table 2.4.

Virus excretion in faeces can occur 2–3 weeks before the onset of symptoms and can last for at least 6 weeks after the onset. Faeces may contain $>10^6$ infectious virus particles per gram. The concentration of the virus in stool declines after jaundice appears, although in some cases virus shedding has been observed for 3 months after the onset of clinical symptoms. Although HAV is mainly shed in faeces, it can also be found in the urine and saliva of infected persons.

The severity of the disease is age dependent, with milder symptoms reported in children than in adults. In children under 6 years of age, asymptomatic infections are common, although severe diarrhoea has been reported in some children. Acute liver failure can result in death in the over-50s, with mortality in this group being around 1–2%. Besides the general symptoms, there are also symptoms characteristic of severe infection, such as sudden vomiting, a tendency to bruise and bleed easily, and drowsiness and confusion. In some cases, atypical manifestation of disease such as acute kidney injury has been reported. Infection during pregnancy is associated with maternal complications and premature labour. There is no evidence of chronic liver disease caused by HAV or for persistent infection with the virus. Complete clinical recovery takes place within 2 months in about 60% of patients, and in almost 100% within 6 months, and most ill persons fully recover with no subsequent associated liver disease.

Infection with HAV leads to the formation of M (IgM) and G (IgG) immunoglobulins. Anti-HAV IgM appears with the symptoms of the disease and usually remains detectable in patients for up to 6 months. Anti-HAV IgG is detectable later during infection and remains in an infected person's serum throughout their life, conferring lifelong immunity to the virus.

Hepatitis E Virus

The discovery of HEV was somewhat bizarre. In 1983, the Soviet physician Mikhail Balayan noted several cases of hepatitis among Soviet soldiers in the Afghanistan war. Although he wanted to continue the investigation in Moscow, refrigeration systems were not available, so he prepared a yoghurt with the faecal samples of nine infected soldiers, drank it, then travelled to Moscow, and waited for symptoms. After 35 days, he developed an acute infection and he started to collect his own faecal samples, in which he found a new virus producing hepatic damage in animals and which could be observed by electron microscopy.

HEV was previously classed as a calicivirus, but now belongs to the genus *Orthohepevirus*, within the family *Hepeviridae*. HEV has now formally been given the taxonomical name *Paslahepevirus balayani* – Pasla, after the animal genera the virus infects, and balayani, after its discoverer (see Box 2.2). The *Hepeviridae* family also contains viruses that infect birds and fish. The avian HEV and the trout HEV share about 50% of the nucleotide sequence identity with the mammalian HEV strains, but have not been associated with any human cases. In recent years, a rat HEV – *Rocahepavirus ratti* – has

Box 2.2 A New Nomenclature for Viruses

In a departure from traditional virus nomenclature, the International Committee for the Taxonomy of Viruses has constructed names for several virus genera and species using a system similar to that used for cell-based organisms, e.g. bacteria, plants, and mammals.

The genus name *Paslahepevirus* is constructed from Primates (P), Artiodactyla (A), Scandentia (S), and Lagomorpha (La). These host groups include species such as humans, pigs, rabbits, and tree shrews.

The genus name *Rocahepavirus* is constructed from Rodentia (Ro) and Carnivora (Ca).

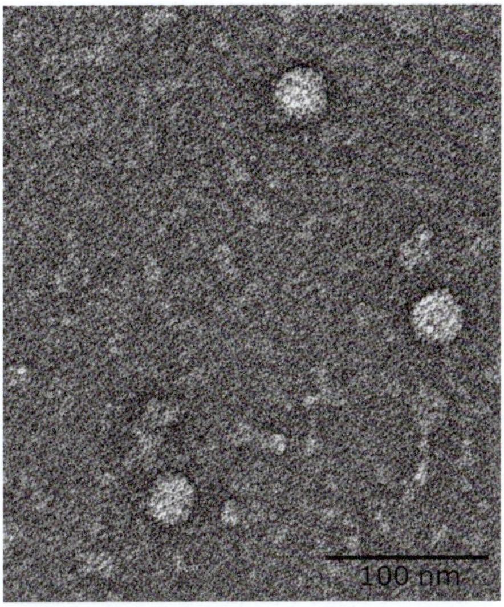

Figure 2.5 Hepatitis E virus particles. Transmission electron micrograph of hepatitis E virus particles from cell culture supernatant. *Source*: A.K. Gadicherla / R. Johne, Federal Institute for Risk Assessment, Berlin, Germany.

been shown to cause disease in humans and could conceivably have food-borne potential (see Chapter 7 on future challenges).

HEV is a single-stranded RNA virus. The particles are approximately 27–34 nm in diameter, with an unenveloped icosahedral capsid (Figure 2.5), enclosing a 7.5-kb linear RNA genome. Its genome has three ORFs: ORF1 encoding a non-structural polyprotein, ORF2 encoding the viral capsid protein, and ORF3 encoding a small regulatory phosphoprotein. Although HEV is classed as unenveloped, quasi-enveloped particles (eHEV) can circulate in the host's bloodstream; however, the viruses which are shed in the faeces are non-enveloped.

There are currently 8 recognised HEV genotypes (gt, see Table 2.5) and several subgenotypes or strains within each. Strains of gt 1 and 2 infect only humans.

Strains of gt 3 and 4 infect humans as well as several wild and farmed animal species including pigs, wild boar, rabbits, and deer. Gt 7 is a rare type, found to infect camels, but there has been a recorded human infection.

The infectious dose of HEV is unknown. After oral entry, the virus passes through the intestinal tract, where it probably reproduces. The virus then passes to the liver and after replication is released into the bile and blood by mechanisms that are not yet fully elucidated. Replication of HEV in the

Table 2.5 HEV genotypes and host species (recognised as of 2025).

Genotype	Host species
1	human
2	human
3	human, pig, goat, wild boar, deer, rabbit, hare
4	human, pig, goat, sheep, cow
5	wild boar
6	wild boar
7	camel, human
8	camel

liver leads to damage in that organ, but the pathogenesis and mechanisms of hepatic injury during infection are not yet clearly understood; immune mechanisms may be involved.

In humans, HEV infections often occur asymptomatically, but after an incubation period ranging from 15 to 60 days, the disease can manifest as an acute viral hepatitis with jaundice. Viral particles can be found in the bile and faeces of the infected person during the late incubation phase and up to 2 weeks after the onset of clinical disease. HEV is excreted via faeces, at up to 10^8 particles excreted per milligram. Excretion occurs for several days before the onset of symptoms. Clinical symptoms of the disease in most cases are very similar to those described during hepatitis A and include anorexia, nausea, jaundice, fever, and abdominal pain. More seriously, infection with gt 1 or 2 can lead to death in 20% of pregnant women. Neurological symptoms such as meningitis/encephalitis and facial paralysis are reported in many cases. Symptoms normally last for 1–4 weeks but can be longer (6 months) in some cases. Chronic infection can occur but is rarely reported. HEV infection appears to be rare in children. As the age of the patient increases, the symptoms can become more severe. Men appear to be more susceptible to HEV infection than do women, for unidentified reasons.

Occasionally or Potentially Foodborne Viruses

In addition to the three virus types just outlined, there are several types of virus which, having a faecal–oral route of transmission, have the potential to be transmitted via contaminated foods (see Table 2.2). Foodborne transmission has been documented for several of these types, albeit in some instances only rarely. For some virus types, foodborne transmission has been suspected but not proven.

Adenovirus

Adenoviruses are members of the family *Adenoviridae*. They are non-enveloped icosahedral viruses with a diameter of approximately 100 nm. The viral genome is double-stranded DNA, approximately 38 kb in length. Adenoviruses infect a wide range of vertebrate animals. Human adenoviruses are currently classified into seven species (A–G). They can cause a variety of diseases, including conjunctivitis and respiratory disease. Many adenovirus strains can replicate in the gastrointestinal tract and thus can be spread via the faecal–oral route. The enteric adenoviruses, types 40 and 41, included in Group F, cause gastroenteritis, particularly in young children, in whom they are the second most important viral pathogen of childhood gastroenteritis after rotavirus. The main symptoms of infection with types 40 and 41 are diarrhoea and fever and may last up to 2 weeks, although recently type F adenoviruses have been reported as the cause of some cases of hepatitis in children.

Astrovirus

Astroviruses are members of the family *Astroviridae*. They are non-enveloped spherical viruses of ~30 nm with a 6.9-kb RNA genome. They can infect a wide range of animal hosts; humans are infected strains of the genus *Mamastrovirus*. Astrovirus gastroenteritis is common in children and also in the elderly and immunocompromised people. The symptoms are similar to, but generally milder than, those of norovirus gastroenteritis. In rare cases, neurological symptoms have been observed. Human astroviruses have been detected in shellfish, and there has been a handful of reports of foodborne astrovirus infection. Zoonotic transmission by bovine astrovirus may also be possible.

Avian Influenza Virus

Influenza viruses belong to the *Orthomyxoviridae* family. They have a wide variety of hosts, including birds and pigs. Influenza viruses are most commonly known as agents of respiratory disease (flu), often serious, in regularly occurring pandemic outbreaks particularly in the winter months. Influenza viruses are enveloped, variable in size, and contain an RNA genome. There are four genera, A–D, based on genetic and antigenic differences, with Influenza A causing the most human cases. Avian influenza virus, a member of group A, is a highly pathogenic virus with the potential for transmission to humans, causing disease which can be fatal in some cases. The major route of avian influenza virus transmission to humans is direct contact with infected birds or their excreta. Another risk factor is contact with carcasses or blood of infected birds during slaughter and poultry meat handling. The virus has been found in imported frozen duck meat and eggs laid by infected chickens; there is, however, no unequivocal evidence that transmission has ever occurred via consumption of contaminated poultry products.

Coronavirus

Members of the *Coronaviridae* family are large (80–160 nm) enveloped viruses, with an RNA genome. Although coronaviruses are mainly associated with respiratory infections, some types can cause gastrointestinal symptoms, particularly in children. Coronaviruses can infect a variety of animals, and zoonotic transmission has been observed in some instances, the cases of human infection being related to close contact with certain animal species. Middle East respiratory syndrome coronavirus (MERS-CoV), SARS-CoV-1 (Severe Acute Respiratory Syndrome Coronavirus), and SARS-CoV-2 (the agent of COVID-19) have been shown to replicate in the human intestine, with high shedding ($>10^8$ per gram) in faeces, and some people reported gastrointestinal symptoms during the pandemic. Several species, including SARS-CoV-2, have been detected in wastewater, although it is not certain whether the virus particles were infectious. Although there are no reports of anyone acquiring COVID-19 by a foodborne route, SARS-CoV-2 has been shown to be able to survive in an infectious state on the surfaces of foods and food packaging materials for up to several days. A role, however minor, for food in COVID-19 epidemiology cannot be comprehensively ruled out.

Enterovirus

Enteroviruses are members of the family *Picornaviridae*. They are non-enveloped RNA viruses, with particle sizes between 25 and 30 nm. Enteroviruses include coxsackievirus, echoviruses, and poliovirus. They can affect the nervous system, causing a variety of symptoms, including encephalitis and meningitis, in all age groups. Gastrointestinal symptoms have also been observed following enterovirus infection.

Enteroviruses have been isolated from shellfish harvested from sewage-contaminated waters. They have also been associated with some outbreaks transmitted by food and water. Poliovirus has now been all but eradicated in most countries through successful vaccination programs, although it can still cause isolated outbreaks and cases in underdeveloped regions of the world.

Nipah Virus

Nipah virus was first discovered in Malaysia, following an outbreak of encephalitis in the village after which the virus was named. Together with Hendra virus, it is a member of the *Henipavirus* genus in the *Paramyxoviridae* family. The particles are enveloped, contain single-stranded RNA, and vary in size between 120 and 500 nm. Infection in humans is manifested by atypical pneumonia and fever, and lethal encephalitis can also occur. Nipah virus is zoonotic, with fruit bats and flying foxes being natural reservoirs. Humans can become infected through consumption of food items that have been in contact with these animals (or intermediate hosts such as pigs, cats, dogs, horses, cows, and goats) or their excretions. In the first recognised outbreak, contact with domestic pigs was the main transmission routes; consumption of fruits which had been partially eaten by bats were probably the vehicles by which pigs became infected with Nipah virus. Consumption of raw palm juice contaminated with bat secretions and/or excrement was probably the cause of human disease in an outbreak which occurred in Bangladesh in 2008. Nipah virus infection is characterised by a high mortality rate of 40–90%, which makes it the most deadly zoonotic virus known.

Parvovirus

Members of the family *Parvoviridae* are small (18–26 nm), round, non-enveloped viruses, with a single-stranded DNA genome of between 4 and 6 kb in length. Human parvoviruses generally cause respiratory disease, but some strains can also cause gastroenteritis. Parvovirus was implicated in a large outbreak (~800 cases) of gastroenteritis due to consumption of contaminated cockles (a marine mollusc) in the United Kingdom. Other foodborne outbreaks have not been recorded.

Rotavirus

Rotaviruses belong to the *Reoviridae* family. The name originates from the Latin word *rota* (wheel), reflecting the virus' wheel-like morphology as seen under the electron microscope. Rotaviruses have a more complex structure than most other enteric viruses: the 75-nm capsid is composed of three protein shells – the core, the inner capsid, and the outer capsid – which form layers surrounding the virus genome. The genome comprises 11 segments of double-stranded RNA. Each genome segment encodes structural or non-structural viral proteins. There are several groups of rotavirus, termed A–J, based on antigenic variations in the VP6 protein which forms the inner capsid. Further classification of rotavirus strains into G (VP7) and P (VP4) genotypes is related to the virus genes which encode outer capsid proteins and depends on the differences in their nucleotide sequences. The VP4 and VP7 proteins play a main role in an initiation of infection, induce protective immunity, and determine the serotype. Functionally, VP7 is a glycoprotein involved in virus adsorption and penetration in the epithelial cells of the host, while VP4 is a spike-like protein responsible for virus infectivity. Genetic changes result in diversity of circulating G and P rotavirus genotypes. The segmented structure of the virus genome favours the occurrence of exchange of viral genome segments between different strains, a phenomenon described as gene reassortment. Reassortment events can take place when a host is infected with different rotavirus strains; the progeny virus particles which emerge from the cell contain a new combination of gene segments and can possess novel antigenic features. Thus, reassortment may result in the formation of novel virus strains, including potentially zoonotic strains.

Most of the strains infecting humans are group A rotaviruses, although infections by other groups, particularly B and C, are observed. The epithelial villi of the small intestine are the target cells for rotavirus infection. Although infection primarily causes gastroenteritis, neurologic signs have also been observed in some cases. Rotavirus is shed via faeces in quantities $\geq 10^{10}$ particles per ml of faecal material, and its presence can be detected up to 1 week after infection. Children under 5 years of age are mainly affected (rotavirus infection is a leading cause of severe gastrointestinal disease in children globally), although rotaviral gastroenteritis can also occur in adults, particularly the elderly.

Rotaviruses also very commonly infect a variety of animals, including domesticated species such as cattle, pigs, and sheep, and it is possible that zoonotic transmission can occur. Although rotavirus is currently not considered as a major foodborne agent, there have been several outbreaks of rotaviral gastroenteritis reported in which contaminated food was suspected as the vehicle. Considering the scale of rotaviral infections, in both humans and animals, it is likely that contamination of foods is quite prevalent, although no surveys have been conducted to confirm this and what, if any, impact this may have remains unknown.

Sapovirus

Sapoviruses (formerly Sapporo-like viruses) are, like norovirus, members of the family *Caliciviridae*. They are non-enveloped icosahedral viruses with a diameter of approximately 34 nm, possessing a 7.5-kb RNA genome. Infection with sapovirus can cause gastroenteritis and is particularly common in children. The symptoms are similar to those of noroviral gastroenteritis.

Foodborne outbreaks of sapoviral gastroenteritis have been recorded, but currently the virus is not considered to be a major foodborne agent.

Tick-Borne Encephalitis Virus

A member of the family *Flaviviridae*, tick-borne encephalitis virus is a 50-nm diameter, enveloped virus with a single-stranded RNA genome, which can cause disease in animals and humans. It is generally transmitted via the bites of ticks. The symptoms include fever, malaise, and nausea and, in serious cases, encephalitis and meningitis.

Foodborne transmission has been known to occur in some countries, when the virus infects dairy cattle and goats, and when milk or milk products from these animals are consumed without prior pasteurisation.

Epidemiology of the Major Foodborne Viruses

Epidemiology is the scientific study of the distribution, determinants, and patterns of health-related events and diseases in populations. It serves as the foundation for public health research and policymaking, providing valuable insights into the risk factors and causes of illnesses, as well as the effectiveness of interventions and prevention measures. By examining the frequency, distribution, and determinants of health outcomes, epidemiologists aim to identify trends and causal relationships to better understand and control diseases.

When it comes to infectious diseases such as those associated with foodborne viruses, epidemiological studies examine the incidence (new cases) and prevalence (total cases) of infections across different populations and geographic locations. This information is crucial for understanding the disease burden, identifying risk factors associated with foodborne virus transmission and developing effective prevention and control strategies. Epidemiologists use a range of study designs and methods, including observational studies and experimental studies to collect data and analyse the relationships between exposures and health outcomes. They also use various data sources, such as surveillance systems, health records, and surveys, to track disease trends and evaluate the impact of interventions.

Epidemiology is critical in identifying and tracking outbreaks of foodborne virus disease. With the use of advanced molecular techniques, researchers can trace the origin of an outbreak, monitor its spread, and evaluate the effectiveness of interventions to control it. By comprehending the epidemiology of foodborne viruses, public health officials and policymakers can make informed decisions regarding resource allocation, preventive measures, and targeted interventions to safeguard and promote public health. Current information regarding the epidemiology of the three major foodborne viruses follows.

Norovirus

Norovirus is the most common cause of gastroenteritis in the world. The virus is mainly transmitted person to person, and it is difficult to calculate the precise number of cases of norovirus gastroenteritis where consumption of contaminated food has been the cause. This is due to several reasons: mainly the disease is not notifiable, the symptoms are regarded as mild, and sufferers seldom report to a general practitioner. However, the WHO has estimated that annually there are 125 million cases of foodborne norovirus, causing around 35,000 deaths. As an illustrative example of this burden, in the United States where noroviruses are the main causative agent of foodborne gastroenteritis, it has been estimated that there are almost 5.5 million annual cases and more than 22,000 hospitalisations, with an associated annual expenditure of around 2 billion dollars and 5000 quality-adjusted life years (QALYs) – that is, within the US population as a whole, there will be 5000 years without perfect health due to the individuals who have had norovirus gastroenteritis.

Norovirus infection is also termed 'winter vomiting disease', since outbreaks occur more frequently in winter. With regard to foodborne norovirus gastroenteritis, however, this seasonality is not evident since outbreaks are reported throughout the year (although an increase in cases can sometimes be observed around certain dates when eating shellfish is more common, e.g. Valentine's Day).

For many years, most cases of norovirus gastroenteritis were caused by genogroup II genotype 4 (GII.4) strains, although GII.17, which emerged in Asia, has now become the dominant strain in many regions.

Noroviruses are generally considered species specific, although human strains have been found in cattle and pigs. No infections with animal strains have been identified in humans, but nevertheless the existence of genogroups and strains with animal hosts suggests that a possibility of zoonotic transmission could exist. In addition, surveillance of circulating strains in the human population has revealed the presence of several rare genotypes, which may represent zoonotic strains. However, zoonotic transmission has not been proven in any case.

Hepatitis A Virus

The global prevalence of HAV is probably underestimated; however, approximately 14 million annual cases of hepatitis A (27,000 deaths) have been estimated to be foodborne. The incidence of hepatitis A can vary due to a variety of factors such as public health infrastructure and water hygiene conditions. Consequently, there are major geographic differences in hepatitis A prevalence. Most infections occur in Africa, Asia, and Central and South America; these are the areas of high endemicity, and in these areas, infections usually occur among children but adults are generally immune. Central and East Europe are considered areas of medium endemicity. North America, Western and Northern Europe, and Australia are regions of low endemicity; this significantly creates a risk of outbreak occurrence, for example, when HAV-contaminated foods are imported from high-endemicity countries. Although cases of acute hepatitis A may peak during spring and summer, this may be due to a variety of factors, and no seasonality of foodborne disease has been definitively observed.

The number of foodborne HAV cases is difficult to estimate because of the virus incubation period taking several weeks from eating the food to onset of symptoms. During this time, the suspected food items may be no longer available for laboratory testing when cases are reported.

Hepatitis E Virus

Hepatitis E caused by gt 1 and 2 strains is endemic in low- and middle-income countries of the world and appears to be predominantly transmitted person to person, although foodborne transmission cannot be entirely ruled out. In high-income regions, infection with HEV gt 1 or 2 is mainly seen in travellers who have returned from endemic regions. However, in recent years, a growing number of infections with HEV gt 3 have been recorded in several high-income countries. These infections have in most cases been involved with consumption of pork or pork products, although meat from other animals such as deer or wild boar has also been implicated in the transmission of infection.

Domestic pigs are a reservoir for HEV gt 3 and gt 4. The prevalence of antibodies to the virus in pig herds (up to 100% in some cases) is high in every country where investigations have been performed. Weaner pigs (young pigs separated from their sow) become infected at 8–15 weeks of age, after the antibody acquired through the sow's colostrum has disappeared. Up to 90% of pigs are infected by 18 weeks of age. The virus infects the pig's liver, but the animal does not show any clinical signs of infection. There is a viraemic phase of infection, in which HEV circulates throughout the pig's blood system and can be carried into muscle tissue; this phase normally lasts for 1–2 weeks but can be longer. Pigs generally go to slaughter at 6 months of age, but some animals can still be infected with HEV at this point. These pigs show no symptoms and cannot be identified at the slaughterhouse during routine inspection of animal health before slaughtering. In this way, HEV can enter the food supply chain.

Hepatitis E due to infection by gt 3 and 4 is now formally recognised as zoonotic. The scale of the zoonosis can be grasped from the results ofa survey of donated human blood, carried out in England between 2012 and 2013, which found evidence (anti-HEV antibodies, and in some samples viral RNA signifying viraemic particles) which implied that 80,000–100,000 infections had occurred in England during the duration of the study. Not all HEV infections progress to clinical symptoms of disease; however, the number of reported clinical cases is growing in several regions. In Europe, the number of cases increased 10-fold between 2005 and 2015. It is likely that cases of the disease are under-reported: in France, one report estimated that the annual number of foodborne hepatitis E cases is around 59,000, which could be around 30 times higher than the annual number of reported cases. As with hepatitis A and norovirus gastroenteritis, no seasonality of foodborne hepatitis E has been observed.

Sources of Quoted Data

Norovirus infectious dose: Atmar, R.L. et al. (2014). Determination of the 50% human infectious dose for Norwalk virus. *Journal of Infectious Diseases* 209: 1016–1022; Teunis, P.F. et al. (2008). Norwalk virus: how infectious is it? *Journal of Medical Virology* 80: 1468–1476.

Norovirus health burden: Food and Agricultural Organisation (FAO) & World Health Organisation (WHO) (2024). Microbiological risk assessment of viruses in foods, part 1: food attribution, analytical methods and indicators. Microbiological Risk Assessment Series number 49 (Rome); Hoffmann, S. et al. (2012). Annual cost of illness and quality-adjusted life year losses in the United States due to 14 foodborne pathogens.

Journal of Food Protection 75: 1292–1302; Scallan Walter, E.J. et al. (2025). Foodborne illness acquired in the United States—major pathogens, 2019. *Emerging Infectious Diseases* 31: 669–677; Food and Agricultural Organisation (FAO) & World Health Organisation (WHO) (2008). Viruses in food: scientific advice to support risk management activities. Microbiological Risk Assessment Series number 13 (Rome).

Hepatitis A health burden: Food and Agricultural Organisation (FAO) & World Health Organisation (WHO) (2024). Microbiological risk assessment of viruses in foods, part 1: food attribution, analytical methods and indicators. Microbiological Risk Assessment Series number 49 (Rome); Kirk, M.D. et al. (2015). World Health Organization estimates of the global and regional disease burden of 22 foodborne bacterial, protozoal, and viral diseases, 2010: a data synthesis. *PLoS Medicine* 12: e1001921.

HEV infection of pigs: Andraud, M. et al. (2014). Early-life hepatitis E infection in pigs: the importance of maternally-derived antibodies. *PLoS One* 9: e105527; Ianiro, G. et al. (2022). Dynamic of hepatitis E virus (HEV) shedding in pigs. *Animals (Basel)* 12: 1063.

England blood donors survey and HEV: Hewitt, P.E. et al. (2014). Hepatitis E virus in blood components: a prevalence and transmission study in southeast England. *The Lancet* 384: 1766–1773.

Annual cases of foodborne hepatitis E in Europe: European Centre for Disease Prevention and Control (ECDC) (2017). Hepatitis E in the EU/EEA, 2005–2015. Stockholm.

Under-reporting of hepatitis E: Van Cauteren, D. et al. (2017). Estimated annual numbers of foodborne pathogen-associated illnesses, hospitalizations, and deaths, France, 2008–2013. *Emerging Infectious Disease* 23: 1486–1492; Colson, P. and Raoult, D. (2017). Autochthonous hepatitis E: a common and fatal but neglected emerging disease in France. *Clinical Microbiology and Infection* 23: 898–899.

Further Reading

Bonita, R. et al. (2006). *Basic Epidemiology*. World Health Organisation.

Centers for Disease Control and Prevention (CDC) (2025). Estimates: Burden of foodborne illness in the United States.

Cook, N. (ed.) (2013). *Food and Waterborne Viruses*. Cambridge: Woodhead Publishing.

European Food Safety Authority (EFSA) (2014). Scientific Opinion on the risk posed by pathogens in food of non-animal origin. Part 2 (Salmonella and Norovirus in leafy greens eaten raw as salads).

European Food Safety Authority (EFSA) (2014). Scientific Opinion on the risk posed by pathogens in food of non-animal origin. Part 2 (Salmonella and Norovirus in berries).

European Food Safety Authority (EFSA) and United Kingdom Food Standards Agency (FSA) (2016). Summary report of joint scientific workshop on foodborne viruses.

European Food Safety Authority (EFSA) (2017). Public health risks associated with hepatitis E virus (HEV) as a food-borne pathogen.

Food and Agricultural Organisation (FAO) & World Health Organisation (WHO) (2008). Viruses in food: scientific advice to support risk management activities. Microbiological Risk Assessment Series (MRA) number 13.

Food and Agricultural Organisation (FAO) & World Health Organisation (WHO) (2024). Microbiological risk assessment of viruses in foods, part 1: food attribution, analytical methods and indicators. Microbiological Risk Assessment Series number 49.

Goyal, S.M. and Cannon, J.L. (eds.) (2016). *Viruses in Foods*, 2e. Springer.

Melhem, N.M. (ed.) (2019). *Norovirus*. Springer.

Nasheri, N. et al. (2019). Foodborne viral outbreaks associated with frozen produce. *Epidemiology and Infection* 147: e291.

Rodríguez-Lázaro, D. et al. (2012). Virus hazards from food, water and other contaminated environments. *FEMS Microbiology Reviews* 36: 786–814.

Ruggeri, F.M., Di Bartolo, I., Ostanello, F., and Trevisani, M. (2013). *Hepatitis E Virus: An Emerging Zoonotic and Foodborne Pathogen*. Springer.

Sánchez, G. (2013). *Hepatitis A Virus in Food: Detection and Inactivation Methods*. Springer.

White, P.A., Netzler, N.E.M., and Hansman, G.S. (eds.) (2017). *Foodborne Viral Pathogens*. CRC Press.

3

How Does Food Become Contaminated?

For many years, food has been considered as a potential source of infectious pathogens, including viruses. There have been many reported outbreaks of foodborne viral disease, several with well-described features identified through epidemiological investigation. Viruses cannot grow in or on foods, but they may be present as a result of direct or indirect contamination by infected humans or animals. Food can be contaminated in two general ways – intrinsically and extrinsically

- **Intrinsic contamination** occurs where the food-producing animals are naturally infected by the virus, with hepatitis E virus (HEV) currently being the primary example.
- **Extrinsic contamination** results when viruses contaminate the foodstuff from an external source (humans or animals) or vehicle (e.g. water).

Although humans can extrinsically contaminate foods at any point along the supply chain from farm to fork, within this book, 'extrinsic' will refer mainly to virus contamination which occurs at primary production or during processing. Contamination further along the supply chain will be more explicitly referred to as being handler associated.

Common food vehicles of virus transmission are those that are eaten raw or only slightly cooked. These include shellfish and fresh produce (Figure 3.1). Fresh produce such as berry fruits and leafy green vegetables can become extrinsically contaminated on the farm during harvesting by infected persons or through contact with contaminated water. Further extrinsic contamination of food items may take place during processing, including cross-contamination via surfaces or utensils. Handling of food by infected food workers (or infected persons in the home) before consumption, e.g. during the preparation of meals, is another risk for extrinsic contamination. Pollution of the water environment, e.g. by sewage, is the main route whereby

Figure 3.1 Fresh produce such as soft berry fruits, leafy greens, stem/bulb vegetables and bivalve shellfish can become contaminated with viruses. *Source:* (Top Left) Engin Akyurt / Pexels, (Bottom-Left) JillWellington / Pixabay / CC0, (Top-Right) Michael Heinrich / Pexels / CC0 and (Bottom-Right) Elle Hughes / Pexels / CC0.

shellfish, being filter-feeding invertebrates, can accumulate viruses in their gut and thereby become extrinsically contaminated.

One of the main differences between enteric viruses and pathogenic foodborne bacteria is host specificity. While many of the major pathogenic foodborne bacteria, such as *Campylobacter* and *Salmonella*, may also infect a variety of animals, most of the enteric viruses associated with foodborne outbreaks are specific for humans. Until recently, it was thought that all foodborne enteric viruses could only originate in humans and therefore their transmission was limited to infected food handlers, cross-contamination of food, and water pollution. However, several recent outbreaks of HEV have shown that zoonotic foodborne virus transmission exists. HEV is widely prevalent in pig herds and can intrinsically contaminate pork products, e.g. salami. There is increasing evidence that strains of animal rotaviruses can infect humans, and they have been isolated from several human cases. The excreta from infected cattle, pigs, and sheep contain large numbers of infectious viruses and when deposited in fields could pass via run-off water into water sources. Therefore, it is possible that zoonotic viruses could, via manure or slurry, contaminate arable land and possibly extrinsically contaminate crops.

Other features of particular relevance to the epidemiology of enteric viral diseases are their low infectious dose (in contrast to bacterial pathogens of food origin such as *Salmonella* or *Campylobacter*) and the very high numbers shed by infected individuals. These two characteristics facilitate potentially very widespread dissemination of viral particles and give foodborne viral pathogens the opportunity to produce outbreaks affecting a large number of people.

Routes of Contamination

As stated earlier, there are two forms of contamination of foods: extrinsic and intrinsic. Various factors influence each form, and the routes associated with them are different.

Extrinsic Contamination

The ultimate source of enteric pathogenic viruses is the internal organs of infected individuals. These viruses are transmitted from their hosts via faeces and vomitus. Thus, direct or indirect contact of foodstuffs with faecal or vomit material is the cause of their contamination with enteric viruses. This may sound highly unpleasant, but it must be remembered that enteric viruses are incredibly small, and their infectious dose is potentially extremely low, and thus only tiny amounts of faeces and vomit can be required to create a risk of infection. Foodborne viruses do not spoil food, and unlike some forms of bacterial or fungal contamination, the appearance and taste of the contaminated foodstuff is never altered by their presence, which will very seldom be accompanied by detectable amounts of faeces or vomit.

The foods most at risk from extrinsic contamination by enteric viruses are those which are extensively handled or exposed to environmental water during production and are subsequently minimally processed or not processed at all, e.g. eaten raw or only lightly cooked. The major types of such foods are berry fruits, leafy vegetables, and bivalve molluscan shellfish.

For berry fruits and leafy vegetables, the main routes of contamination are contact with sewage-contaminated water and with the hands of infected food handlers. Primary production, pre- and during harvest, is the phase in the food supply chain at which these routes are most significant. Figure 3.2 depicts the main routes of extrinsic contamination of food by viruses. The virus enters the sewage system and can potentially contaminate various

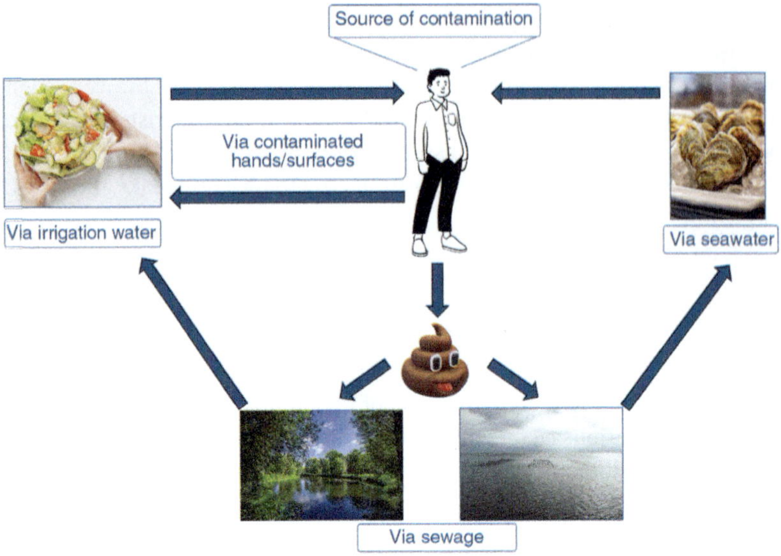

Figure 3.2 The major routes of extrinsic enteric virus transmission. *Source:* Polina Tankilevitch / Pexels (Via seawater)Filipp Romanovski / Pexels / CC0, (Via sewage, left) Nikolaj Erema / Pexels / CC0 and (Via sewage, right) Mingche Lee / Pexels / CC0.

water sources such as rivers, lakes, and the sea. The contaminated water may then be used for irrigation, potentially leading to contamination of crops such as lettuce, strawberries, and raspberries. Once the affected produce is harvested, it can then enter the food supply chain and ultimately be consumed by humans, causing illness.

Sewage treatment is not fully effective at removing viruses, and viruses can be found in high numbers (e.g. approximately 10^3 virus particles per litre) in sewage discharges. Treated (and on occasion raw) sewage can be discharged into rivers, estuaries, and seawater. Enteric viruses can survive for long periods in contaminated water, especially at low temperatures (see Chapter 5). Berry fruits and leafy green vegetables have a considerable requirement for water during their production, and irrigation is required almost constantly during their growth. Irrigation can be performed in two main ways – by drip or using sprinklers. In many modern production sites, drip irrigation is the preferred mode; however, sprinkler (overhead) irrigation is also widely used. Water used for irrigating fresh produce is obtained from a variety of sources including well water, borehole water, surface water, reservoirs, and the potable domestic supply. A high percentage of farms will take their irrigation water from surface water sources, e.g. rivers. In the United Kingdom in 2006, 50% of vegetable growers stated that they harvest some leafy crops within 24 hours of the last irrigation, and this practice is likely widespread. Enteric viruses have been detected in irrigation water in some studies. Viruses can survive for several days on field crops, and also on foods post-harvest (see Chapter 5). If sewage-contaminated water is used for irrigation of crops, for application of pesticides, or for washing of harvested produce, then viruses can become located on the produce and potentially remain infectious through to the time of consumption. It has been estimated that if 100 l of irrigation water is contaminated with 5 infectious HAV particles, then one in every 10,000 lettuces irrigated with this water will cause an infection.

It has been speculated that viruses in irrigation waters could become internalised in the tissues of the crop plants, via uptake through the roots. Some studies have shown that there may be a potential for this; however, their experimental conditions were very artificial, e.g. the plant roots were exposed to enormous numbers of virus particles, and therefore – although it may not be entirely impossible – the likelihood of internalisation occurring naturally is probably very low.

The excreta from livestock may contain large numbers of infectious HEV and rotavirus particles. Viruses in excreta deposited in fields could pass via run-off water into fresh waters such as rivers or lakes. Also, excreta from cattle and pigs can be stored as manure and slurry, then spread onto land as fertiliser. In many countries, there is a round-the-year application of animal wastes. This could mediate a continuous introduction of HEV and rotaviruses into the environment, potentially resulting in the contamination of water courses and, potentially, food crops.

Bivalve shellfish (e.g. oysters, mussels, clams, cockles, and scallops) are filter feeders that process large amounts of seawater to obtain their food. They can accumulate a wide variety of micro-organisms including, if present, human pathogens. Virus contamination of shellfish harvesting areas can result from continuous discharges of treated municipal sewage (Figure 3.3), periodic (intermittent) untreated discharges (combined sewer overflows, storm tank overflows), leaks from ageing or poorly maintained sewerage infrastructure, smaller discharges from individual properties e.g. septic tanks, and discharges from boats.

A food commodity which has been rising in popularity over recent years is edible seaweed. With the increase in popularity of sushi and Japanese restaurants, there is a greatly increased consumption of seaweeds. Edible seaweed is now considered a significant part of the food industry: approximately 85% of harvested seaweed is for human consumption. In Asian countries such as Japan, Korea, and China, seaweed is a traditional food.

Figure 3.3 Sewage outflows into shellfish harvesting areas or into rivers, waterways, and lakes which are sources of irrigation water, can be the source of viral contamination. *Source:* Helio Dilolwa / Pexels / CC0.

It is well known for its use in sushi, salads, and dried snacks. In other countries, such as the United States, where it has not been traditionally consumed, its popularity has risen to the point where it is now being considered as a whole food, mainly because of its nutritional benefits. However, being harvested from seawater, it also means this product could be at risk from virus contamination, especially if the coastal waters where it is grown are contaminated by sewage. There have been several outbreaks of norovirus associated with the consumption of seaweed.

Poor hand hygiene in food handlers is a significant factor in contamination of foods. Viruses can be transferred from contaminated hands or fingertips onto food items or food preparation surfaces, particularly if moisture is present (Figure 3.4). A UK Food Standards Agency survey of food service workers found that only 35% understood the risks of pathogen transmission through touching food with bare hands. This study also found that 22% of handlers interviewed admitted that if they were busy, they found it difficult to perform effective hand hygiene. Furthermore, only 3% recognised the potential problem of asymptomatic food handlers. As discussed in Chapter 2, approximately 16% of norovirus-infected persons show no symptoms. If such persons do not carry out effective hand hygiene, they will pose a substantial risk of contaminating foods. In a study in Japan, for example, norovirus-asymptomatic handlers were identified in several settings

Figure 3.4 Many soft fruits are still harvested by hand. If poor hand hygiene is present, this can be the cause of transfer of viruses from humans to foods.
Source: Iwona Kozyra and Artur Rzeżutka.

(e.g. restaurants, care centres, hotels, and hospitals) where outbreaks had occurred, and it was concluded that the situation was widespread.

Foodborne transmission is a significant factor in norovirus outbreaks on cruise ships. Buffets, self-serve stations, and shared food handling areas create opportunities for cross-contamination. Infected food handlers who fail to practice proper hygiene can introduce norovirus into meals, affecting large numbers of passengers.

Handling of fresh produce such as berry fruits and leafy vegetables can occur at any stage in their supply chains – during harvest, during processing, at point of sale, during food service, or during the preparation of meals in the home. Handling of these foodstuffs by virus-infected persons is a risk factor for transmission of enteric viruses. Poor personal hygiene, e.g. not washing hands either through lack of concern or lack of available facilities among food handlers, can be a contributing factor in outbreaks of norovirus gastroenteritis. Other potential factors which should be considered are major failures in food safety culture, i.e. accidental or deliberate contamination of foods with faecal material, either maliciously or through ignorance of the consequences.

Contamination of food can also occur in processing or preparation environments through contamination or cross-contamination of surfaces or utensils. Studies have shown that enteric viruses can be transferred from contaminated foods to hands, and from hands to surfaces and vice versa. Utensils, such as knives used for slicing vegetables, have been shown to be capable of transferring viruses to additional produce items after processing contaminated items.

Intrinsic Contamination

Animals used as food sources can be hosts for various viruses that are, or are potentially, foodborne (see Chapter 2). Currently, HEV is the most significant zoonotic foodborne virus.

Consumption of raw or undercooked pork products has been identified as a major risk factor for HEV infection, with the possibility that several thousand people become infected each year. HEV RNA sequences (implying intact virus particles) have been found in various pig tissues and organs at slaughter, and in pork products at retail sale around the world (see Figure 3.5).

Wild boars, and other animals such as deer and rabbits, can also be hosts for HEV. In many countries, these animals are extensively hunted for food. If animals are infected when they are killed, foods prepared from them are at risk of containing HEV, and hepatitis E has been observed in people who have consumed such food and subsequently developed symptoms.

Cross-contamination from butchering utensils may also spread HEV from infected animals to products from non-infected animals.

HEV has also been detected in mussels, shellfish, and other bivalve molluscs; they will have become contaminated via run-off water from farms or agricultural land. Furthermore, HEV sequences have been detected in samples of berries and vegetables, with irrigation water suspected as the source of the contamination. Thus, the virus can contaminate foods extrinsically as well as intrinsically.

Outbreak Descriptions

There are two basic ways whereby outbreaks of foodborne viral disease can be detected in the community. The first relies on identifying patterns of illnesses among cases who have reported eating the same food at similar times, or eating together at a specific location, e.g. at a celebration in a restaurant. An epidemiological investigation is then conducted, in which various types of information (e.g. the symptoms, and their onset and duration) from each case is gathered by the investigators to try to identify the food vehicle, and possibly the source and route of contamination.

The second way involves observing an increase in the number of disease cases above what is considered usual in a community or region, or seeing cases occurring in a cluster. During the subsequent epidemiological investigation, affected persons are interviewed to try to identify if there were common reasons for why they became infected, e.g. consumption of identical food items.

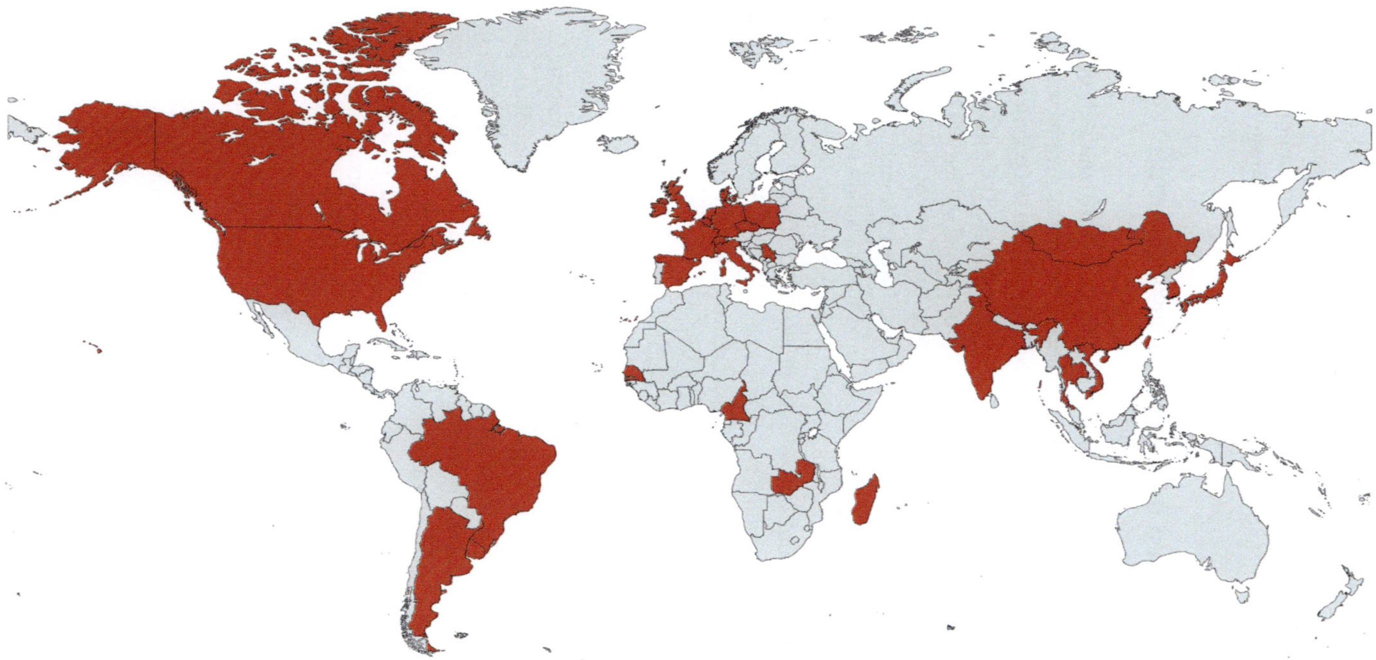

Figure 3.5 Countries (in red) where HEV RNA sequences have been detected in pig tissues and organs at slaughter, and in pork products at retail sale (as of 2025).

As has been mentioned, there have been many recorded outbreaks of viral disease due to consumption of food. Studying the features of such outbreaks can be useful to illustrate some of the issues surrounding foodborne virus contamination, and some example descriptions are given here.

Outbreaks Caused by Extrinsic Contamination

In May 2013, public health authorities in both Germany and Italy observed that the number of hepatitis A cases reported in various regions was higher than the number which was normally expected within the population. HAV genotype 1 was identified in the patients' samples. The epidemiological investigation found that the infected persons had all consumed frozen berries sold commercially. The outbreak spread to other countries in Europe, including Ireland and the Netherlands, and lasted through to 2014. HAV was detected in batches of the frozen mixed berries that had been implicated as the vehicle of infection. Within these batches, blackberries produced in Bulgaria and redcurrants produced in Poland were the most consistently present fruits. The European Food Safety Authority conducted an investigation, and the berry lots were traced back to processors and to some of the farms where they were harvested. The investigation did not, however, determine how thc berries became contaminated. This illustrates the fact that, in most of the recorded outbreaks where fresh produce has been implicated as having transmitted foodborne viruses, how the produce actually became contaminated has not been definitively demonstrated.

A further example is provided by the large outbreak of norovirus gastroenteritis which took place in Germany in 2012. Initially, over 4000 cases were reported from 5 regions of the country; finally, over 11,000 cases were recorded. The epidemiological investigation linked the cases to the consumption of frozen strawberries, which were traced to a shipment of 22 tonnes which had been imported from China. Analysts discovered three different norovirus genotypes (GI.9, GII.13, and GII.16) in samples taken from this shipment, indicating that the viruses came from several persons, since individuals are normally infected by one strain at a time. This might imply that the strawberries were contaminated via sewage contamination. German experts went on a fact-finding mission to China and visited the fields where the berries had been harvested but could not identify any obvious source of contamination, although it was noted that sewage pipes ran nearby.

An outbreak of hepatitis A with subsequent legal consequences occurred in March 1997, in the United States, when a large cluster of hepatitis A cases was noted among schoolchildren in Michigan. There was a strong association between illness and consumption of food items containing frozen strawberries. These items had been eaten mainly during Valentines' Day lunches the previous month (the onset of symptoms several weeks after consumption of contaminated food being typical for hepatitis A; see Chapter 2). The strawberries were traced to a Californian food processor, who had imported them from Mexico. They had been picked in April 1996, then sent to the processor in San Diego, who froze the batches in April/May. The fruit was stored frozen before being sold to the US Department of Agriculture (USDA), which was buying foodstuffs as part of a school lunch program. This strongly illustrates the prolonged survival capacity of viruses, particularly when frozen (see further Chapter 5). The processor was charged with several violations of US federal law, which pertained to the fact that the USDA required that any foodstuff purchased for the school lunch program should have been sourced within the United States, but the processor had acquired the berries abroad. One of the directors was given a suspended prison sentence, and the company was ordered to bear the cost of disposal of all the frozen berries from the implicated shipment, a quantity which totalled thousands of tonnes. The company subsequently filed for bankruptcy. Some small quantities of the frozen strawberries remained in researchers' freezers, and, finally, after standardised methods became available, HAV was detected in one of the remaining batches in 2017.

Sometimes, it can be surmised that infected handlers have been the source of contamination in fresh produce-associated outbreaks. In Scotland in 1988, five persons who ate a raspberry dessert together became ill with hepatitis A. The dessert was prepared using fruit bought at a Tayside farm. An investigation revealed that several of the raspberry pickers had been ill, with signs of jaundice. One of the harvesters was confirmed as having hepatitis A. In this case, contamination of the raspberries by the handlers was most likely.

In November 1988, an outbreak of hepatitis A involving 43 cases was reported in Ohio, United States. All of these persons had eaten meals containing salad onions at a local restaurant. There was no sign of infection in any of the restaurant's employees. The US Food and Drug Administration (FDA) traced the implicated onions back to one farm in California and to two in Mexico. It was observed that on these farms, at least three workers were required to prepare onions, one each performing peeling, trimming, and bundling of the vegetables. Once again, contamination by infected handlers was the likely source of the virus, although as it was also noted that the onions were shipped packed in ice and therefore a role of contaminated water could not be ruled out.

In many outbreaks which have occurred in catering premises, it is clear that food handlers preparing the food have been the cause. A good example is an outbreak recorded in Belgium in 2004, in which 269 individuals were infected by HAV. Raw beef was identified as the vehicle of virus transmission. An individual had manually prepared beef meat before its distribution to local butcheries. The food handler had contracted hepatitis A infection and was working with bare hands in the days preceding his symptomatic illness. Blood and serum samples collected from infected persons and from the food handler were HAV-positive. Subsequent analysis confirmed that the infections were caused by the same strain of HAV which the food handler was infected with. Although meat samples were not tested for the presence of HAV (there was at the time no effective method for this), there was enough evidence to state that consumption of raw beef was a probable source of infection. As HAV is not zoonotic, it is very likely that the lack of observance of personal hand hygiene by the food handler allowed HAV to enter the food chain. Another example of a possible lapse in hand hygiene is an outbreak of norovirus gastroenteritis affecting nearly 200 people that occurred in four hospitals in Cardiff, Wales in 1994. The source of this outbreak was a food handler who prepared salads in a central kitchen serving each hospital. At home, she had been looking after her baby who had a gastrointestinal illness. One day she vomited in the kitchen and was immediately excluded from work. The same day the first cases appeared among the hospital patients; all were linked to the consumption of salad dishes. But these cases could not have been linked to the vomiting episode, as they occurred within the 12-hour norovirus incubation period. They must have been linked to the salad items which had been prepared the previous day. The food handler had contaminated these items either directly (and presymptomatically), or indirectly via vomit or faeces which her baby had shed on her hands. Either way, poor hand hygiene was probably implicated.

Bivalve shellfish can readily become contaminated by enteric viruses, if the water in which they live is contaminated, e.g. by proximity of sewage outflows, or the run-off from septic tanks or agricultural land. Consumption of virus-contaminated shellfish has been the cause of many outbreaks of viral disease. An example is an outbreak of hepatitis A which occurred in New South Wales, Australia in early 1997, where oysters were implicated as the vehicle of infection of several hundred people. It was possible to trace back the oysters from the point of purchase to the growing beds. An environmental investigation revealed several potential sources of pollution of the lake where the growing beds were situated, with sewage or human faeces. This pollution could have originated from a town upriver of the lake where the septic tanks of several households drained into storm-water channels, from picnic and camping areas in proximity to the lake which had no toilet facilities, or from several vessels sailing on the lake which had inadequate waste disposal systems. In some samples of oysters harvested at the same time as those implicated in the outbreak, some other enteric virus types (enteroviruses, adenoviruses, and reoviruses) were detected together with HAV, suggesting multiple contamination sources. The investigators could not definitively identify any major pollution event, but some weeks prior to the outbreak there had been exceptionally high rainfall in the area, which may have caused wide dispersal of the viruses. HAV could be detected in oyster samples for more than 2 months after the presumed contamination event, which indicates the scale of persistence of HAV in contaminated oysters and the limited effectiveness of conventional treatments for virus removal from shellfish (see Chapter 6).

Another example occurred in the United States in 2005. Here, the outbreak-related oysters were farmed and harvested from two areas in the Gulf of Mexico. The oysters were shipped to several restaurants where they were eaten raw. None of the cases reported any risk factors related to infection other than food consumption. Oysters harvested from the same areas were collected for laboratory testing. It was confirmed that all cases were due to infection by a single strain classified as HAV genotype IA; an identical strain was detected in the oysters. This outbreak originated from a common source, and it is likely that this source contaminated the oysters before distribution since employees involved in shipment had minimal physical contact with the shellfish and none reported hepatitis symptoms. All implicated oysters had been harvested from approved areas remote from any fixed source of sewage discharge. Contamination of the shellfish beds by harvesters was unlikely as they met all hygienic requirements. The probable sources of contamination were illegal human waste disposal from recreational boats within legal harvest areas or illegal harvesting. The shellfish-growing waters are shallow, and contamination even from a single defecation event could have reached the beds.

In Japan in 2017, over 2000 people became ill with norovirus after consuming dried, shredded seaweed (also called 'nori'). Investigators in Japan found that the contamination had originated at a seaweed production plant. It was the cause of seven foodborne norovirus outbreaks: Tokyo, as well as four remote areas of Japan, was hit by the outbreaks. The investigators found norovirus GII.17 in stool samples from patients. This particular strain is a recent variant which emerged in Asia in 2014/2015 and is now becoming dominant (see Chapter 2). The investigation showed that the seaweed was served before the onset of all the outbreaks, and norovirus was detected in dishes containing the 'nori' seaweed. In each of the seven outbreaks, the nori was served as a topping or as part of the meal. The nori was found to have come from the same food processing company in each case.

Outbreaks Caused by Intrinsic Contamination

Due to the prevalence of the virus in domestic pigs, the potential scale of zoonotic exposure to HEV is substantial (see Chapter 2), although this has not been reflected in the incidence of reported outbreaks of hepatitis E implicated to the consumption of raw, minimally processed, or improperly cooked pork products (Figure 3.6). This may be due to the circumstance that such products can be widely sold at retail and consequently cases may be sporadic, in contrast to easily distinguishable outbreaks which can occur after consumption of contaminated food at specific events or locations. Three cases of hepatitis E in Japan were linked to the consumption of grilled pork meat and entrails in a barbeque restaurant in Japan in 2012, and in France in 2012, 17 individuals became infected after eating a roast piglet stuffed with the animal's raw liver. Also, in France, in 2010, three cases of hepatitis E were associated with the consumption of figatellu, a sausage made using pigs' liver and often consumed raw.

Several reports of HEV infection have been linked to eating meat from wild animals which have been hunted. In 2003, four people in Japan presented to a hospital with hepatitis, and HEV was detected in their serum. It was initially thought that the index patient had acquired the infection through foreign travel, but examination of the patients' histories revealed

Figure 3.6 Pork products such as salami and raw pork sausage can become contaminated with HEV. *Source:* RDNE Stock project / Pexels and Domenico Scaglione / Pexels / CC0.

that they had all eaten the meat from two Japanese Sika deer caught in the wild. The meat had been eaten raw, sushi-style, several times during the 3 weeks preceding the onset of the patients' disease symptoms. Some left-over portions of the deer meat had been kept frozen, and upon analysis HEV was detected in the meat from one deer. The virus sequences from the deer were identical to the virus isolated from the patients. These patients belonged to two families who shared occasional meals, but only those members who ate notable amounts of the contaminated deer meat became infected. In 2015, an HIV-infected patient in Cordoba, Spain, was diagnosed with an acute hepatitis E infection. The patient was negative for HEV on serological testing, but the virus was detected in his blood. Eight members of his family were also found to be infected, but without showing symptoms. The patient's family traditionally hunted wild boar for consumption, and the whole family consumed stewed, roasted, or grilled boar meat during the previous month. Two pieces of boar meat provided by them were analysed, and HEV gt3 was observed. The viruses in the affected family members and the meat samples were identical. The prevalence of HEV among wild boars in the same hunting area was also evaluated, and the same strain was detected in all tested samples.

This review of outbreak features strongly reveals the overall challenges of food virological safety – the scale of some outbreaks, the difficulty in pinpointing how the food became contaminated, and the risks of some consumption practices. The specific challenges surrounding detection of viruses in foods will be discussed in the next chapters.

Sources of Quoted Data

Time of harvesting of crops in UK: Tyrell, S. et al. (2006). Microbiological water quality requirements for salad irrigation in the United Kingdom. *Journal of Food Protection* 69: 2029–2035.

Estimation of HAV contamination of lettuces via irrigation: Stine, S.W. et al. (2005). Application of microbial risk assessment to the development of standards for enteric pathogens in water used to irrigate fresh produce. *Journal of Food Protection* 68: 913–918.

Production of edible seaweed: de Souza Celente, G. et al. (2023). Seaweed as an alternative protein source: prospective protein extraction technologies. *Innovative Food Science & Emerging Technologies* 86: 103374.

UK Food Standards Agency survey of food service workers: Social Science Research Unit, Food Standards Agency (2017). Food handlers and norovirus transmission: social science insights.

HAV outbreak, Europe, 2013: Severi, E. et al. (2015). Large and prolonged food-borne multistate hepatitis A outbreak in Europe associated with consumption of frozen berries, 2013 to 2014. *Euro Surveillance* 20: 21192.

NoV outbreak, Germany, 2012: Bernard, H. et al. (2014). Large multistate outbreak of norovirus gastroenteritis associated with frozen strawberries, Germany, 2012. *Euro Surveillance* 19: 20719; Mäde, D. et al. (2013). Detection and typing of norovirus from frozen strawberries involved in a large-scale gastroenteritis outbreak in Germany. *Food and Environmental Virology* 5: 162–168.

Hepatitis A outbreak, USA, 1997: Hutin, Y.J. et al. (1999). A multistate, foodborne outbreak of hepatitis A. *New England Journal of Medicine* 340: 595–602; Cook, N. et al. (2021). Detection of hepatitis A virus in strawberries implicated in an outbreak in the USA in 1997. *Food and Environmental Virology* 13: 421–422.

Hepatitis A outbreak, Scotland, 1988: Ramsay, C.N. and Upton, P.A. (1989). Hepatitis A and frozen raspberries. *Lancet* 1 (8628): 43–44.

Hepatitis A outbreak, USA, 1988: Dentinger, C.M. et al. (2001). An outbreak of hepatitis A associated with green onions. *Journal of Infectious Diseases* 183: 1273–1276.

Hepatitis A outbreak, Belgium, 2004: Robesyn, E. (2004). An outbreak of hepatitis A associated with the consumption of raw beef. *Journal of Clinical Virology* 44: 207–210.

Norovirus outbreak, Wales, 1994: Lo, S.V. et al. (1994). The role of the pre-symptomatic food handler in a common source outbreak of food-borne SRSV gastroenteritis in a group of hospitals. *Epidemiology and Infection* 113: 513–521.

Hepatitis A outbreak, Australia, 1997: Conaty, S. et al. (2000). Hepatitis A in New South Wales, Australia from consumption of oysters: the first reported outbreak. *Epidemiology and Infection* 124: 121–130.

Hepatitis A outbreak, USA, 2005: Bialek, S.R. et al. (2007). Use of molecular epidemiology to confirm a multistate outbreak of hepatitis A caused by consumption of oysters. *Clinical Infectious Diseases* 44: 838–840.

Norovirus outbreak, Japan, 2017: Kusumi, E. et al. (2017). Multiple norovirus outbreaks due to shredded, dried, laver seaweed in Japan. *Infection Control and Hospital Epidemiology* 38: 885–886.

Hepatitis E outbreak, Japan, 2012: Miyashita, K. et al. (2012). Three cases of acute or fulminant hepatitis E caused by ingestion of pork meat and entrails in Hokkaido, Japan: zoonotic food-borne transmission of hepatitis E virus and public health concerns. *Hepatology Research* 42: 870–878.

Hepatitis E outbreak, France, 2012: Guillois, Y. et al. (2015). High proportion of asymptomatic infections in an outbreak of hepatitis E associated with a spit-roasted piglet, France, 2013. *Clinical Infectious Diseases* 62: 351–357.

Hepatitis E outbreak, France, 2010: Colson, P. et al. (2010). Pig liver sausage as a source of hepatitis E virus transmission to humans. *The Journal of Infectious Diseases* 202: 825–834.

Hepatitis E outbreak, Japan, 2003: Tei, S. et al. (2003). Zoonotic transmission of hepatitis E virus from deer to human beings. *Lancet* 362: 371–373.

Hepatitis E outbreak, Spain, 2015: Rivero-Juarez, A. et al. (2017). Familial hepatitis E outbreak linked to wild boar meat consumption. *Zoonoses and Public Health* 64: 561–565.

Further Reading

Elviss, N.C. (2022). Detection of norovirus from the commercial food premises and food handlers. *Journal of Food Protection* 85: 2001–2009.

European Food Safety Authority (2017). Public health risks associated with hepatitis E virus (HEV) as a food-borne pathogen. *EFSA Journal* 15: 4886.

Food and Agricultural Organisation (FAO) & World Health Organisation (WHO) (2024). Microbiological risk assessment of viruses in foods, part 1: food attribution, analytical methods and indicators. Microbiological Risk Assessment Series, No. 49.

Food and Agricultural Organisation (FAO) & World Health Organisation (WHO) (2008). Viruses in food: scientific advice to support risk management activities. Microbiological Risk Assessment Series, No. 13.

Lin, D. (2025). Viral transmission in seafood systems: strategies for control and emerging challenges. *Journal of Food Safety* 45: e12981.

Nasheri, N., Vester, A., and Petronella, N. (2019). Foodborne viral outbreaks associated with frozen produce. *Epidemiology and Infection* 147: e291.

Nemes, K. (2023). Hepatitis A virus and hepatitis E virus as food. *Microorganisms* 11: 1712.

Nguyen, R. (2023). Sources of microbial contamination in vegetable and animal-sourced food. *Food Safety and Hygiene* 10: 44–52.

Olaimat, A.N. (2024). Common and potential emerging foodborne viruses. *Life* 14: 190.

Rodríguez-Lázaro, D. et al. (2012). Virus hazards from food, water and other contaminated environments. *FEMS Microbiology Reviews* 36: 786–814.

Yang, Z. et al. (2017). Internalization and dissemination of human norovirus and Tulane virus in fresh produce is plant dependent. *Food Microbiology* 69: 25–32.

4

Detection

The detection of viruses in foods poses a greater challenge than that of most foodborne bacteria. Enteric viruses are sub-microscopically small: for example, hepatitis A virus (HAV) and norovirus are only around thirty millionths of a millimetre in diameter. Due to their probable low infectious dose, they need to be present as contaminants in a food in only very small numbers to constitute a risk to health. They do not change the appearance or sensorial qualities of food, and their presence is undetectable by sight or smell. They are incapable of growth in food, and unlike foodborne bacteria, their numbers in a food sample cannot be increased by enrichment in nutrient media in the analytical laboratory. For these reasons, complex methods must be applied by the analyst in order to ascertain whether a foodstuff is contaminated by viruses. These methods are primarily composed of two stages: sample treatment and assay. Sample treatment is a multi-step process aimed at removing viruses from a food sample and concentrating them for delivery to the assay, which can be in various formats but is commonly nucleic acid amplification based on the polymerase chain reaction (PCR). Although several approaches are being proposed, determining whether viruses isolated from foods are infectious or not is highly challenging – and unlikely to be realistically routinely achievable using current methodologies. Nonetheless, useful methods for virus detection in a variety of foodstuffs are available and have been internationally standardised.

Successful virus detection in food depends not only on the method used but also on the sampling strategy which is applied, e.g. the number of samples that need to be analysed and from which part of a batch these samples are taken. Currently, there are no specific guidelines addressing food sampling for virus analysis, and existing sampling recommendations do not necessarily meet virological requirements. Furthermore, differences in the sample size and its initial treatment can have an influence on the detection limit of the method used. Fresh produce may be contaminated by a small number of viruses, and they could be heterogeneously distributed on the surfaces of the produce. These factors significantly hamper proper sampling by making it difficult to select a representative food sample for testing. Sampling will not be dealt with further within this book.

Sample Treatment

Sample treatment for foodborne virus detection is generally based on four principal steps: (1) removal of viruses from a foodstuff into a liquid suspension, (2) removal of any remaining food substances from the virus suspension, (3) concentration of the suspended viruses for delivery to a detection assay, and (4) extraction of nucleic acids from the concentrated viruses. Within each step, variations can exist in application and approach depending on the food matrix/virus combination, and this is reflected in methods which are available for use by the analyst.

Sample Size

When food is tested for viruses, a reasonably large portion should be analysed to increase the possibility that contamination can be detected. The convention in food microbiology is generally to analyse 25 g of food, and some methods for virus detection likewise use that amount: for example, the International Standards Organisation (ISO) standard methods for berries and leafy green vegetables (see Box 4.2 and 4.3 later in this chapter).

But 25 g in real terms may represent only a very small sample of a food portion and could be unrealistic with regard to the actual amount consumed; for example, a single strawberry can weigh 25 g, and it is very likely that more than one berry is consumed in a meal. Analysing a larger sample size (e.g. the average portion size that is consumed in a meal) could be more valuable for risk analysis, as the number of microorganisms detected in the sample would reflect the number that would be ingested and therefore that the consumer would be exposed to. However, in foodborne virus detection, large samples sizes can pose a serious challenge. There are a few reasons for this. Bacterial detection is mainly facilitated by increasing the number of bacterial cells in a sample by enrichment (i.e. by adding the sample to a nutrient-rich medium and allowing the bacteria to multiply to detectable levels) – but with viruses, this cannot be performed because adding a food sample to a cell culture system enhancing virus growth would damage the cultured cells and the viruses would not be able to multiply. Moreover, detection of viruses is usually performed using a nucleic acid assay which operates using microlitre volumes, and therefore the sample must ostensibly be highly concentrated to allow delivery of the viruses or their extracted nucleic acid to the assay. The larger the size of the sample, the less easy it becomes to concentrate it, and the greater the risk that interfering substances could be co-concentrated and delivered along with the target (see later in this chapter).

With shellfish, because most virus contamination is found in the digestive organs of these animals, the majority of protocols used for virus extraction analyse the digestive glands, after their removal from the shellfish bodies. The glands from several shellfish are pooled to make up a test portion. This approach allows increased probability of virus detection in a food lot and eliminates excess food matrix.

Methods for the detection of hepatitis E virus (HEV) in meat and meat products and liver and liver products, although they have been used successfully in some studies, are still in the experimental stage (at the time of writing this book, an ISO Working Group was actively developing an international standard based on RT-PCR). The samples taken in these methods are small – in most methods to date, only a few milligrams – but as there is a high probability of a large number of viruses being present if the sample comes from an infected animal, it may not be absolutely necessary to analyse large sample sizes if animal-derived products are the major ingredient. However, if a processed food in which animal products are only a minor ingredient (e.g. some varieties of sausage or paté) is to be analysed, determining a suitable sample size is a considerable challenge.

Release of Viruses from the Foodstuff

When viruses contaminate foodstuffs such as berry fruit and leafy green vegetables, the greatest likelihood is that they will be located on the surfaces of the foodstuff, rather than internalised within the plant cells or tissues. Electrostatic and hydrophobic attractions can occur between virus and food surface and need to be broken to mediate release of the particles. An effective means of breaking these interactions is increasing the pH to >7, and therefore washing of the foodstuff with an alkaline solution or buffer is commonly employed to elute virus particles from food samples. With some fragile foods, e.g. berry fruits, a gentle washing procedure in the buffer is performed to avoid breaking up the food (see Figure 4.1). With more solid foodstuffs such as salad vegetables, a more vigorous approach can be followed, such as shaking or vortexing. The addition of protein, such as beef extract or soya protein, to the eluant can increase the effectiveness of removal by competing with virus capsids for protein-binding sites on the food surface and thus hindering re-attachment. Soft fruit contains pectin, which can form a jelly during the sample treatment process; virus particles become encased in this jelly and are almost impossible to separate from it. The addition of pectinase to the eluant minimises jelly formation. The end result of this

Figure 4.1 Gentle rolling of strawberries and lettuce samples in an alkaline buffer to remove viruses from food surfaces. *Source:* Photo by Martin D'Agostino.

Figure 4.2 Preparation of oyster samples for virus analysis. *Source:* Photo by Martin D'Agostino.

step is that virus particles are in suspension in the buffer, although this suspension will still also contain some food-derived substances.

With shellfish (see Figure 4.2), viruses are released from the digestive gland tissue by homogenisation, which can be performed with addition of, e.g. a proteinase enzyme to mediate further dissolution of the tissues. The resulting homogenate contains liberated virus particles. Similar principles are applied in the analysis of meat and liver, and associated products, for HEV, where viruses are liberated from samples through homogenisation with different chemical solvents or buffers.

Removal of Food Substances

After elution of viruses, it is necessary to remove the gross food solids, e.g. intact berries or leaves. Generally, this is done by using a filter bag through which the suspension is decanted to a new container. Next, smaller suspended food substances need to be removed from the virus suspension. During this step, with berry fruits and leafy vegetables, the suspension should be kept at high pH to prevent the re-adsorption of viruses to food particles. The main approach to food substance removal is low-speed centrifugation of the suspension, i.e. at 10,000 × g (see Figure 4.3). The addition of a cationic flocculant can enhance the removal of suspended food solids. Finally, viruses are left in a suspension which is free as far as possible from food particulates. However, the volume of this suspension may be extremely large (e.g. millilitres) relative to the volume (i.e. microlitres) of sample that will be analysed at the final assay step, and therefore subsequent concentration of the virus particles is needed.

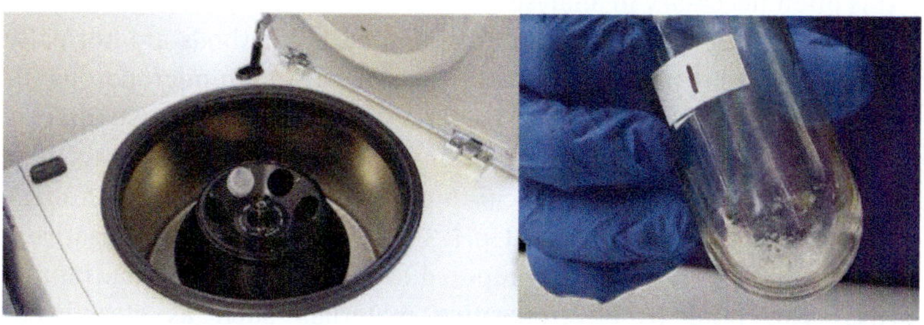

Figure 4.3 A refrigerated benchtop centrifuge used to separate solid food particles from suspension and a close-up of a typical pellet after centrifugation. *Source:* Photo by Martin D'Agostino.

Concentration of Viruses

This step applies to the analysis of berry fruits and leafy greens. With shellfish, and in the current methods for HEV detection, the suspension from the previous step is taken forward to the nucleic acid extraction step without further concentration.

As well as the need to reduce the volume of virus suspension, the possibility of low virus numbers in the original sample must be taken into account. Different strategies have been proposed for the concentration of viruses, including precipitation, filtration, centrifugation, ultracentrifugation, and immunocapture. All have advantages as well as disadvantages.

A widely used method for virus concentration is precipitation of virus particles out of suspension with polyethylene glycol (PEG). PEG reduces protein solubility, and its addition can make the virus particles bind to each other and to any food-derived proteins remaining in the suspension. This suspension needs to be adjusted to a neutral pH to enhance flocculation. A short slow-speed centrifugation will pellet the virus particles, which is then re-suspended in a small volume of liquid (the concentrate). PEG is inexpensive and easy to use and can facilitate good recovery of virus. As it may interfere with the final assay step, however, it is necessary to remove it from the concentrate, normally by extraction with a chloroform–butanol mixture.

Ultracentrifugation is very rapid but requires expensive instrumentation. It can be used to sediment viruses out of suspension by centrifugation at around 230,000 × g. This method is highly effective in concentrating all viruses. However, if the eluate from the previous steps contains high amounts of food substances, the amount of debris in the concentrate will be considerable and could interfere with the subsequent steps of analysis.

It is often necessary to analyse water which is used in food production settings. In these circumstances, filtration-based methods are normally used. Viruses may be concentrated by passage through microporous filters, similar to those used to trap bacteria. The pore size of these filters is much greater than the diameter of any enteric virus, but due to their proteinaceous capsids, viruses can bind to the filter material through electrochemical interactions. At pH 3–6, they will bind to negatively charged filters such as those composed of nitrocellulose; at neutral pH, viruses can bind to electropositive materials such as glass wool (see Figure 4.4). Elution from the filter is performed using a pH >7 buffer eluant (see Figure 4.5). Subsequently, PEG is added to the eluate to

Figure 4.4 Filtration of environmental water through a glass wool filter setup. *Source:* Iwona Kozyra and Artur Rzeżutka.

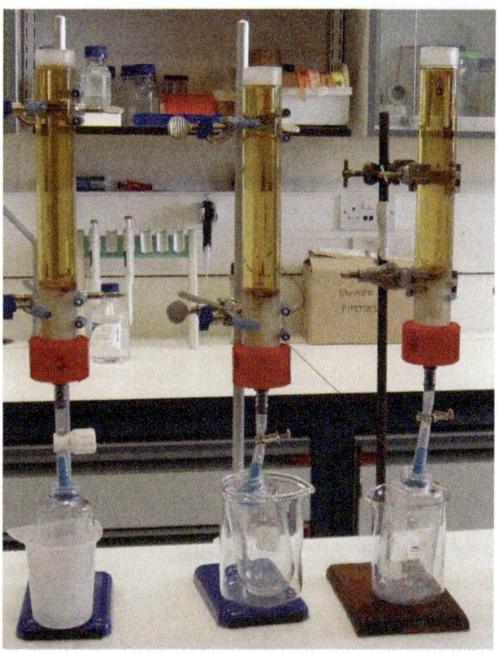

Figure 4.5 Passage of eluant to detach viruses from glass wool filters. *Source:* Photo by Martin D'Agostino.

flocculate the viruses, and thus virus concentration is achieved. An advantage of using filters is that a high volume of suspension can be handled; however, a disadvantage is the possibility of clogging of the filter by suspended particulates, e.g. if surface water used for irrigation is analysed.

In ultrafiltration-based concentration, pores in a membrane that are around 10 nm in size permit passage of liquids and low-molecular-mass particles but exclude viruses and macromolecules. An advantage of ultrafiltration is that there is no prior need to adjust the pH of the virus suspension. Again, however, clogging can be a major problem. This can be overcome to some extent in tangential flow ultrafiltration systems, where a continuous flow is passed across the membrane, and liquid passes through the pores while larger-sized particles are swept across them. The system circulates the suspension, and viruses are retained in a diminishing volume of liquid until the retention volume of the system is reached.

The end result of the concentration step is that the viruses that were in the original food sample are now in suspension in a small-volume (around 0.5–1.0 ml) concentrate.

Nucleic Acid Extraction

Although there are several assay approaches for detection of foodborne viruses (see the following section), the most effective and conventionally used are based on amplification of viral nucleic acid. Prior to their application, a further step of sample treatment is necessary – nucleic acids must be purified from the concentrated virus suspension. Highly efficient commercial kits are available for nucleic acid purification, some specifically marketed for use on viruses. Most of these kits use the principle of breaking the virus capsid through alkaline lysis, then binding the released genomic material to silica and washing away impurities with a low pH buffer. Impurity removal is of major importance when analysing complex matrices like foodstuffs, which contain substances that are highly inhibitory to nucleic acid amplification assays and can be extracted and concentrated along with viruses. Finally, a neutral buffer is used to elute the purified nucleic acids into a small volume (generally around 50–100 µl) of solution. The end result is a purified extract containing the nucleic acid from viruses in the food sample. This represents a considerable concentration of the sample: the original, e.g. a 25-g foodstuff is now a 50-to-100-µl nucleic acid extract, which is ready for testing by a nucleic acid amplification assay.

Detection Assay

There are four principal assay approaches which can be employed for detection of foodborne viruses: electron microscopy, infection of cultured mammalian cells, immunodiagnostics, and nucleic acid amplification. Each has advantages and disadvantages, as will be described in the following sections.

Electron Microscopy

Electron microscopy (EM) is a valuable technique which has been extensively used in classical virology, to allow direct virus identification in the investigated sample. The analyst can actually 'see' the virus particles and can take images of them for further identification (these images can be quite visually pleasing, see Figures 4.6 and 4.7, and others within this book). However, despite this advantage, EM is rarely used for the detection of foodborne viruses as it has limited sensitivity and specificity and can be very time-consuming.

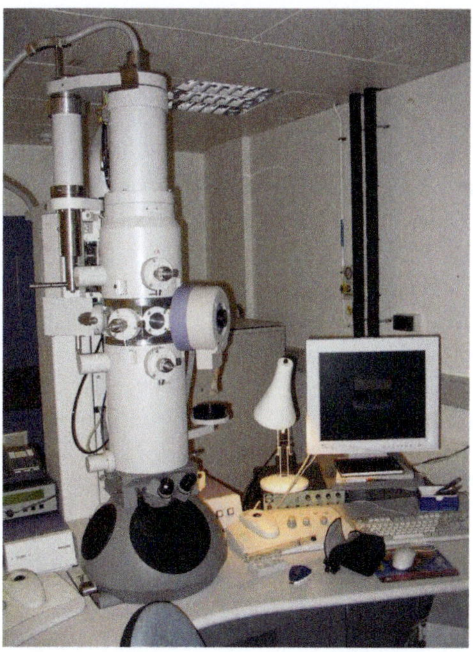

Figure 4.6 Transmission electron microscope. *Source:* David J. Morgan / Wikimedia Commons / CC BY-SA 2.0.

(a) (b)

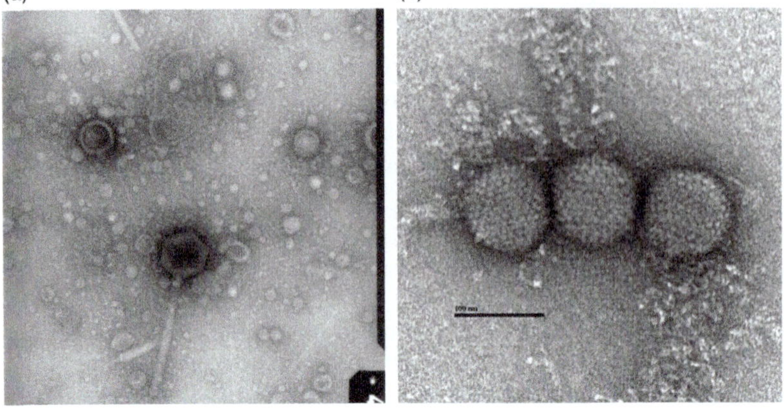

Figure 4.7 Electron micrographs of viruses. (a) A bacteriophage; (b) rotavirus particles. *Source:* Photo by Nigel Cook.

Cell Culture

The outstanding advantage of cell culture is that it allows unambiguous determination of virus infectivity. In this approach, a culture of mammalian cells is inoculated with a virus suspension and monitored for signs of infection. For some virus types, e.g. enteroviruses, infection of cultured cells can produce characteristic damage to susceptible cells, described as cytopathic effect (CPE) formation (Figure 4.8). However, this approach is not suitable for use in food sample analysis, largely because the majority of the strains of the major foodborne virus types are not cultivable or do not produce a visible CPE.

As some enteric viruses will replicate in cell culture without producing a prominent CPE, to improve the sensitivity of the assay, especially for non-cytopathogenic virus types, it can be combined with nucleic acid amplification: after a few days of incubation, samples are taken and checked for an increase in the nucleic acid amplification signal, indicating an increase in virus particle number through replication. This is the basis of some newly developed assays for hitherto difficult-to-culture agents such as HEV and norovirus. See Box 4.1.

Immunodiagnostics

This approach, where antibodies against the virus target are used to capture the virus particles, and with the detection signal produced using dyes or radioactive labelling, can confer great specificity, i.e. the virus targets of

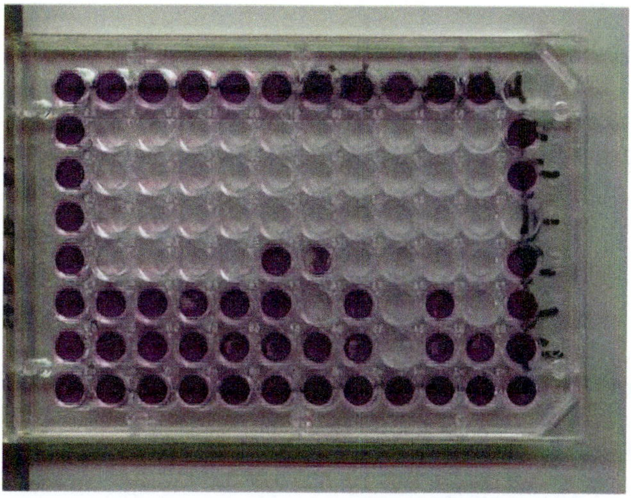

Figure 4.8 A cell culture showing CPE after infection with poliovirus. Here, the cells were grown in a microtiter plate, then successive dilutions of a poliovirus suspension were added to each row. The number of wells showing CPE can be used to calculate the number of infectious virus particles, similar to a Most Probable Number method as used to estimate bacterial viable numbers. *Source:* Photo by Nigel Cook.

Box 4.1 Evaluation of Norovirus Infectivity Using Human Intestinal Enteroids (HIEs)

For many years, virologists had been attempting to cultivate human norovirus in vitro, but no method could be successfully reproduced. In 2016, however, researchers at Baylor College of Medicine, Houston, Texas, reported a successful method for norovirus replication using enteroid cells from the human intestine,* with the addition of bile to mimic natural conditions. In this approach, the enteroids are grown in monolayers, to which norovirus is added. Although CPE can be observed in some instances, norovirus genome equivalents are measured by RT-PCR at addition and after certain time periods. Increase in genome equivalents indicates norovirus replication. Although this approach is technically challenging and costly to perform, it is becoming more accessible to the research community and is being increasingly used to generate data on norovirus survival and to determine the efficacy of disinfection procedures.

*Ettayebi K, et al. (2016). Replication of human noroviruses in stem cell-derived human enteroids. *Science* 353: 1387–1393.

interest are detected but any others present are not. There are immunological tests such as enzymatic immunoassay, radioimmunoassay or enzyme-linked immunosorbent assay (ELISA), and many are commercially available for testing clinical samples for the main enteric viruses. However, their analytical sensitivity is too poor for effective testing of food samples, in which the number of virus targets can be low.

Nucleic Acid Amplification

Nucleic acid amplification (NAA) has the potential for exquisite sensitivity, theoretically being capable of amplifying one target molecule in a single reaction. Because NAA targets specific nucleic acid sequences, detection of virus from strain to genus level can be mediated by amplification of sequences in characteristic genomic regions. This is achieved through careful oligonucleotide primer design, and in this way NAA assays have been developed for strains of all the known foodborne viruses.

The NAA assay which has been predominantly employed in analysis of foods for viruses is the PCR. This employs a heat-stable DNA polymerase enzyme (often sourced from thermophilic bacteria, such as those found in hot springs), and two short nucleic acid molecules (oligonucleotide primers, usually just called 'primers'), containing nucleotide sequences matching those of the intended target, in a cyclic reaction to produce exponentially increasing copies (termed 'amplicons') of a DNA target sequence (see Figure 4.9). The reaction proceeds in two or three phases (denaturation, hybridisation, and elongation), together comprising one PCR cycle. During the denaturation step (at 93–96°C), the double-stranded (ds) DNA 'melts', opening up to two single-stranded (ss) DNA molecules. During the hybridisation step (usually at 55–65°C), the primers anneal to complementary regions of ssDNA (see Figure 4.10), and finally the extension phase is carried out across the target sequence by the DNA polymerase. In many PCRs, the hybridisation and elongation steps are combined and are performed at one temperature. After each PCR cycle, the newly synthesised DNA strands then serve as templates for the next cycle. Several cycles (normally between 30 and 50) are performed, in a process called 'thermocycling'.

In the early development of PCR, detection of amplicons was based on gel electrophoresis. In this approach, primers are designed to bind to target sequences to produce amplicons of definitive sizes after thermocycling. When the reaction is completed, a coloured dye is added to the completed reaction solution, which is then added to wells formed in a slab of agarose (the gel). The gel is made by dissolving agarose powder in a buffer solution, before pouring the mixture to form the slab. The slab is placed in an

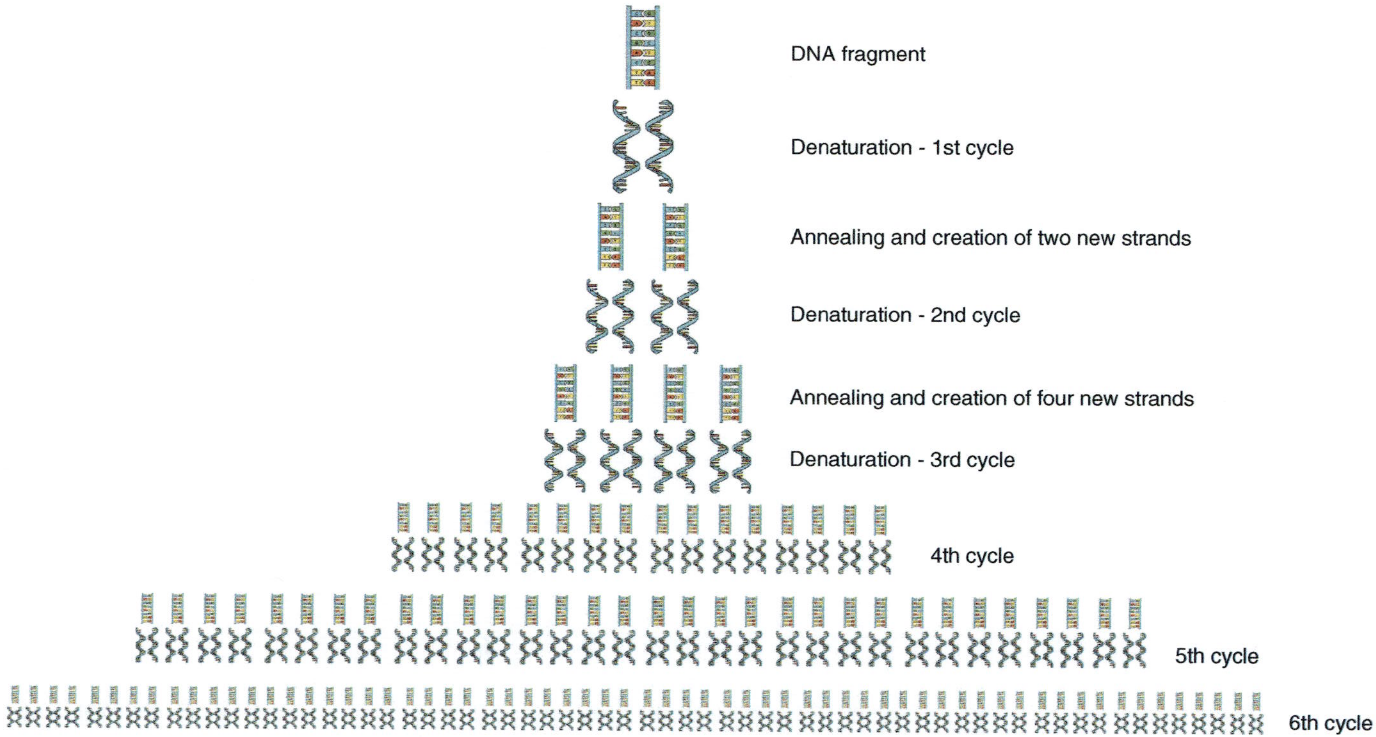

Figure 4.9 The PCR amplification process, which typically comprises 40–50 cycles. *Source:* Martin D'Agostino.

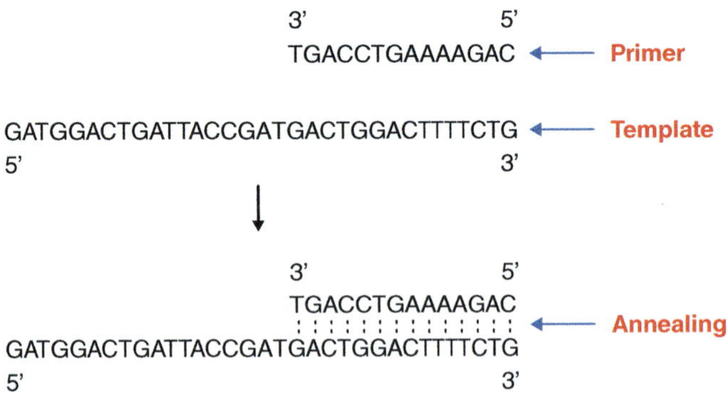

Figure 4.10 Primer binding during the PCR process.

electrophoresis tank and covered in buffer. An intercalating fluorescent dye, very commonly ethidium bromide, is added to the gel before pouring and to the covering buffer. The molecules of this dye insert into ds DNA and thus bind to the amplicons. An electric current is then applied, with the negative terminal behind the well end of the gel and the positive terminal beyond the foot of the gel. As DNA is negatively charged, amplicons will migrate through the gel towards the positive terminal; their progress impeded to a greater or lesser extent by the gel, according to their size. Meanwhile, the dye molecules intercalate into the amplicons. The coloured dye also migrates towards the positive terminal, and when it reaches near the end of the gel, the current is switched off and the gel removed. The gel is placed on an ultraviolet lamp (a transilluminator) and when the lamp is switched on the intercalating dye fluoresces, with very high intensity where its molecules are stacked within the amplicons. The amplicons can thus be visualised directly and photographed (see Figure 4.11). If all is correct, the amplicon size will match the distance between the primer binding sites. This approach is termed 'conventional PCR' and is still in use in many analytical laboratories. A second round of PCR can be performed to further confirm identity: This is often done using a set of primers which match sequences within those targeted by the first PCR and is termed 'nested PCR' (nPCR; see Figure 4.12).

As mentioned in Chapter 2, the major foodborne viruses possess RNA genomes. For detection of RNA, a step prior to PCR is required to transform the RNA into an amplifiable DNA template. This is performed using the enzyme reverse transcriptase, which synthesises a complementary strand of DNA (cDNA) along the length of the target RNA sequence (see Figure 4.13). This cDNA is denatured in the first PCR cycle, a primer binds to the DNA

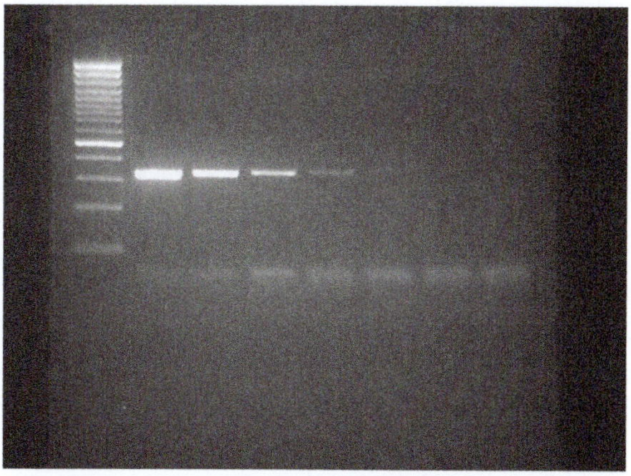

Figure 4.11 Conventional gel-based detection of PCR amplicons. The lane to the left of the image is a ladder of calibration standards increasing in 100 base pair increments; the other lanes contain a dilution series of a target amplicon. The faint bands below 100 bp are primer dimers, caused by some primers binding non-specifically to each other. *Source:* Photo by Martin D'Agostino.

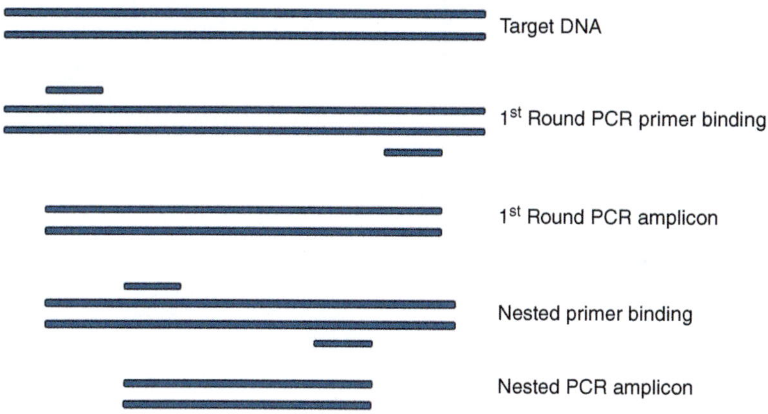

Figure 4.12 The nested PCR process.

strand and extended, and the resulting double-stranded DNA molecule is amplified during subsequent thermocycling. This hybrid reaction is termed 'reverse transcription PCR' (RT-PCR), and this technique will be the one which is specifically referred to henceforward in this chapter.

Nowadays, the most commonly used NAA method is real-time PCR, where amplification of target sequences is monitored through the increasing fluorescence resulting from incorporation of dyes, or

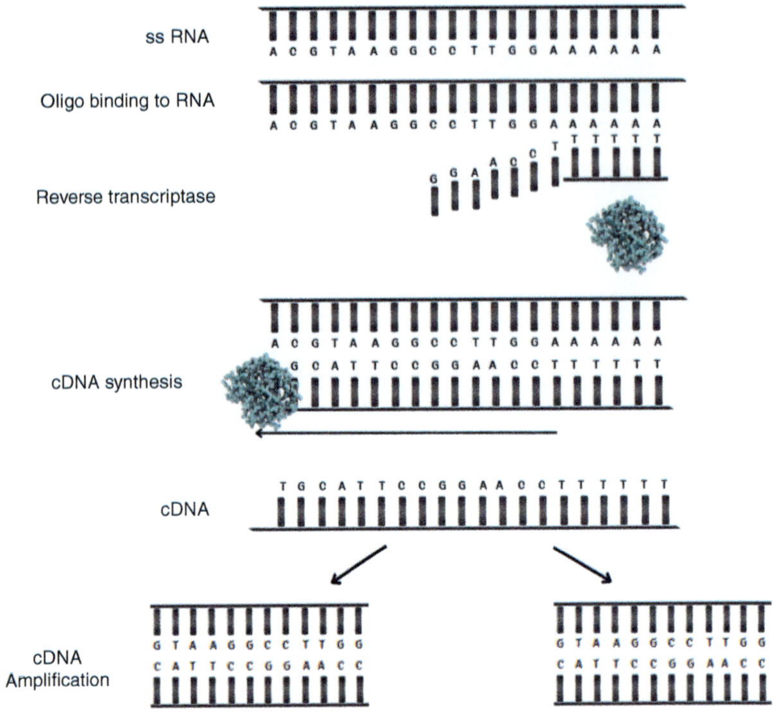

Figure 4.13 The RT-PCR process. *Source:* Designed by Martin D'Agostino, using Canva.

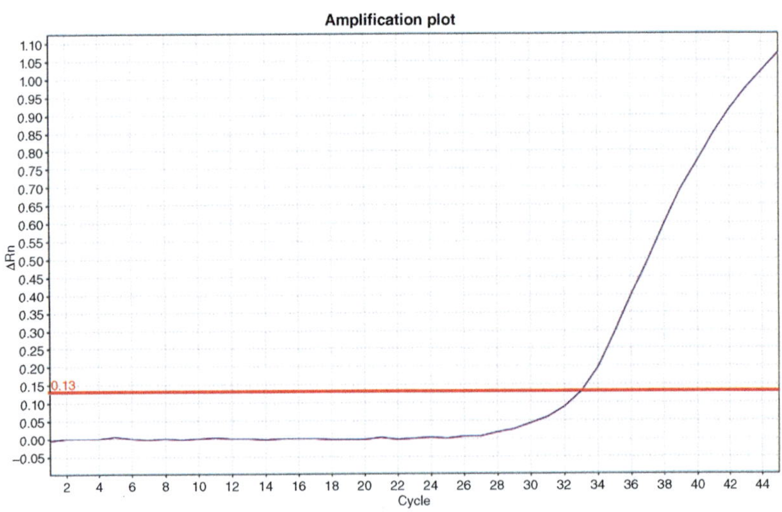

Figure 4.14 Real-time PCR amplification plot showing positive detection of a target (its fluorescence signal crossing the threshold).

attachment of dye-labelled probes over a specific time, and displayed on a monitor or screen linked to the instrument. Unlike conventional PCR where the reaction contents must be removed from the tube prior to amplicon visualisation, in real-time PCR, the signal is monitored within a closed tube; this considerably reduces the potential for contamination of a laboratory with amplicons. Again, unlike conventional PCR, positive detection of the target is determined not by visualisation of amplicons, but by noting when the target signal crosses a pre-set threshold fluorescence intensity (see Figure 4.14).

Quantitation

NAA-based assays can be performed qualitatively or quantitatively. Qualitative NAA-based assays show only whether the target is present or absent in the reaction, as judged by the production or non-production of a target signal. Real-time PCR can be adapted through the use of calibration to act quantitatively. Calibration is generally performed by using a solution of pre-quantified target genome copies, diluted to produce a range of quantities (calibration standards). The RT-PCR is performed, and the cycle number where the signal crosses the pre-set threshold fluorescence intensity (the Ct value) is correlated with the number of genome copies to produce a calibration curve (Figures 4.15 and 4.16). Then, when analysing a real-life sample in which the number of targets is unknown, the Ct value is read against the calibration curve, and the number of original targets in the sample can be ascertained (Figure 4.15; most modern real-time PCR instruments have software which will perform this automatically).

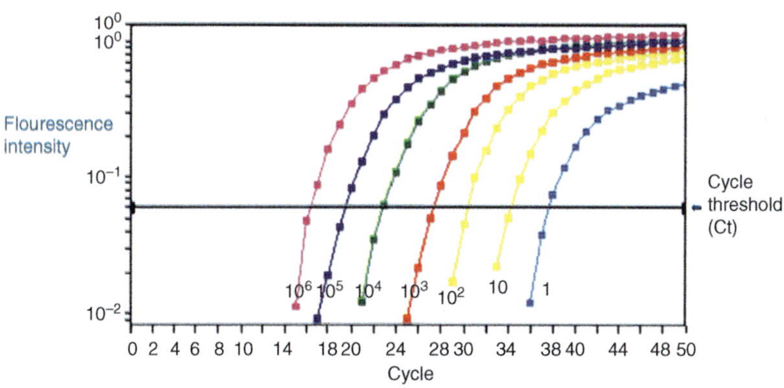

Figure 4.15 An example of an amplification plot of calibration standards using real-time PCR. *Source:* David Rodriguez-Lazaro.

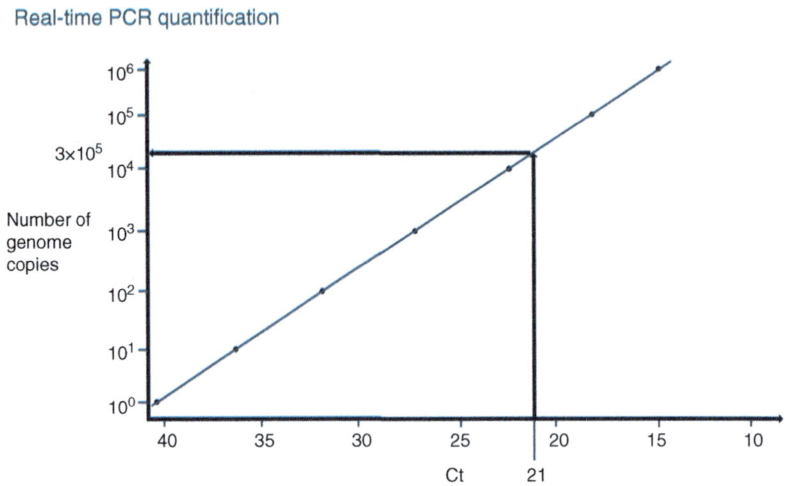

Figure 4.16 A calibration curve constructed from the data in Figure 4.15. An example of a test resulting in a Ct value of 21 is shown; using the curve gives a reading of approximately 3×10^5 genome copies. *Source:* Nigel Cook.

Quantitative detection of RNA viruses such as norovirus ideally requires that the calibration standards be composed of RNA, to fully reflect the RT-PCR process and facilitate reliable quantitation. Several published assays have used ds plasmid DNA constructs as calibration standards, but as the reverse transcription step is not accounted for, and PCR and reverse transcription have different reaction efficiencies (i.e. PCR is more efficient at replicating DNA than RT is at transcribing RNA), this will result in an underestimation of the numbers of virus target (Figure 4.17).

Quantitative results obtained by a properly calibrated (i.e. using RNA standards) RT-PCR assay can be interpreted as representing the number of viral genomes present at the start of the reaction; those obtained by an improperly calibrated assay (i.e. using DNA) cannot. In this book, RT-PCR will be used for non-quantitative assays, RT-qPCR for assays which have been quantified using DNA, and qRT-PCR for assays which have been quantified using RNA standards.

Quantitation can also be performed in digital format. This is where a dilution series of a virus suspension is made and aliquots of each dilution analysed by RT-PCR; the end-point dilution(s) where target signal(s) is observed is used to calculate the number of particles in the original suspension, in the same manner as a most-probable number technique.

Foods are complex analytical matrices and can contain substances which can easily interfere with a detection method. This is particularly the case with

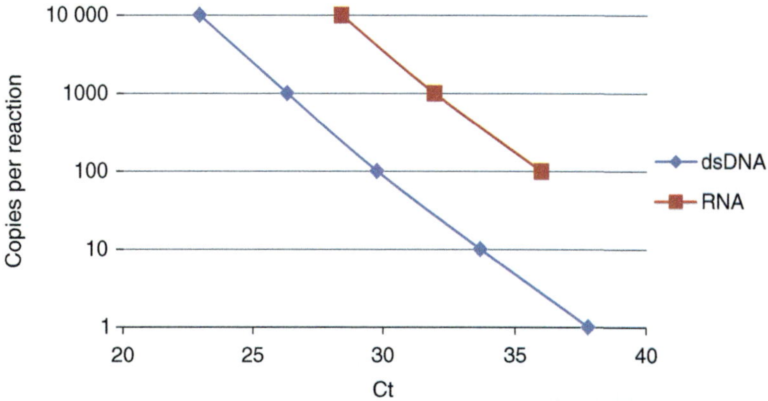

	dsDNA calibration standard	ssRNA calibration standard
Ct value	32	32
Genome copies	30	1000

Figure 4.17 Real-time PCR calibration plots using DNA and RNA calibration standards, illustrating the underestimation of target numbers when using DNA compared to RNA standards (Cook and D'Agostino, unpublished).

RT-PCR-based methods, where inhibitory substances co-extracted with viral nucleic acid can prevent amplification and thus the appearance of a positive signal, consequently leading to false-negative interpretation of results. Matrix-induced reaction failure can, in some instances, be combatted by the use of facilitators – substances added to the reaction mix which prevent inhibitors from interfering with the reaction and ensure the production of a signal even if inhibitors are present.

As well as false negatives, PCR-based methods can, especially due to their very high sensitivity, be susceptible to false-positive results, if the assay has not been designed correctly or if cross-contamination has occurred. When the target has been amplified successfully, the resulting reaction solution contains massive quantities of amplicons, such that even a few picolitres could contain sufficient numbers to be detectable by a subsequent amplification. If a new batch of samples becomes contaminated by previously produced amplicons, this could result in false-positive interpretation of the analysis. As mentioned earlier, in real-time PCR, the whole process from amplification to signal detection is performed in a closed tube, reducing the chance of amplicons being released into

Box 4.2 The ISO Standard Methods

There are two ISO standards for detection of hepatitis A virus and norovirus in foods:

ISO 15216-1:2017
Microbiology of the food chain – Horizontal method for determination of hepatitis A virus and norovirus using real-time RT-PCR – Part 1: Method for quantification
ISO 15216-2:2019
Microbiology of the food chain – Horizontal method for determination of hepatitis A virus and norovirus using real-time RT-PCR – Part 2: Method for detection

Each standard contains methods for the analysis of four food supply chain matrices: (1) berry fruit, leafy green, and bulb vegetables, (2) bivalve shellfish, (3) bottled water, and (4) food (contact) surfaces. The sample treatment methods described in each standard are identical.

The RT-PCR assays described in each standard are identical in their basics (primers/probes, etc.). In ISO 15216-2:2019, the assay is used for presence/absence determination. In ISO 15216-2:2017, calibration by plasmid standards is described, to produce a quantitative RT-PCR assay (RTqPCR).

Both parts describe the use of an internal process control (IPC); a genetically modified virus (Mengo virus strain MC_0) is detailed as an example in the standards, although other types may be used, e.g. murine norovirus (MNV). In these standards, the IPC is termed 'SPC' or sample process control. The standard stipulates that in order to report a sample as truly negative, the efficiency of recovery (using the IPC to calculate this) must be at least 1%.

Both parts recommend the use of external amplification controls (EACs), although alternatives are permitted if demonstrated to be as effective.

the laboratory environment. Moreover, if care is taken during reaction preparation, e.g. by preparing the reaction mix in a separate laboratory, the chance of contamination should be further greatly reduced. However, for reasons of space, many laboratories perform sample preparation in the same lab as the PCR thermocycler is located, and therefore contamination of sample by target amplicons can never be ruled out, although it is strongly advised to physically separate the sample preparation area from the PCR assay area in order to minimise the contamination potential. A technique which can further reduce the possibility of false-positive results via amplicon contamination is the use of the enzyme uracil n-glycosylase (UNG). UNG cleaves DNA strands containing uracil, a

Box 4.3 Food and Drug Administration (FDA) Bacteriological Analytical Manual (BAM) Compendium Standard

There is a standard in the Food and Drug Administration (FDA) Bacteriological Analytical Manual (BAM) Compendium for concentration, extraction, and detection of enteric viruses from food.

The standard contains methods for the analysis of four food supply chain matrices: (1) green onion and leafy greens, (2) fresh and frozen soft fruits, (3) molluscan shellfish, and (4) scallops and finfish. For food matrixes in categories (1) and (2), sample treatment is based on ultra-centrifugation. The standard recommends a process control virus (murine norovirus), and an internal amplification control.

The RT-PCR assays are qualitative.

nucleic acid which can bond with adenosine. If uracil is incorporated in the mastermix (mixture of PCR reagents), resulting amplicons can be degraded in subsequent reactions by pre-treatment with UNG in the next mastermix.

There are now international standard methods for detection of hepatitis A and norovirus in various foodstuffs, published by ISO (see Box 4.2). These are becoming widely used and will be the 'gold standard' for some time. Some national standard methods for virus detection also exist, e.g. in the United States (see Box 4.3).

Interpretation of Results

If monitoring of food supply chains for viruses as part of a food safety program or an epidemiological investigation is to be performed effectively, then it is vitally necessary that the reliability of the analytical results can be verified. Foods such as leafy green vegetables, shellfish, and soft fruit are complex matrices and are difficult to analyse and can furthermore contain substances which can inhibit RT-PCR. The use of facilitators or techniques used for the prevention of cross-over contamination cannot completely eliminate the chance of reaction failure or contamination from occurring. It is essential, therefore, that verification of results includes recognition of analyses where the method has failed to perform correctly, as the presence of a virus in a sample may go undetected through a false-negative interpretation of the absence of a signal. Therefore, during routine monitoring of foodstuffs with an RT-PCR-based method, an appropriate suite of controls must be used to prevent false-negative or false-positive interpretations of the results and to verify that the method has performed correctly (Table 4.1).

Table 4.1 Required controls in the analysis of food samples for viruses.[a]

Control	Definition	Purpose
Internal Process Control (IPC)	A non-target virus added in a defined number to each sample (including the NPC) at as early a step as possible in sample treatment and preferably upon receipt in the laboratory.	To verify that sample treatment has been successful. Also, an estimate of target recovery efficiency can be made by comparing the number of IPC viruses detected with the number that was added.
Negative Process Control (NPC)	Target-free matrix (of the same weight or volume as that of a matrix sample extract), which is subjected to the sample treatment procedure. An IPC is added to the NPC.	To verify that the reagents and equipment used in sample treatment are free of target virus or target amplicons.
Negative Extraction Control (NEC)	Water or buffer without matrix (but of the same volume as that of a matrix sample extract) which is subjected to the nucleic acid extraction procedure.	To verify that the reagents and equipment used in the NA extraction procedure are free of target virus or target amplicons.
Negative Amplification Control (NAC)	Water or buffer (of the same volume as that of a matrix sample aliquot) which is subjected to RT-PCR.	To verify that the reagents and plasticware used for RT-PCR are free of target virus or target amplicons.
Positive Amplification Control (PAC)	A reaction containing the target nucleic acid in a defined amount or copy number.	To verify that the RT-PCR is capable of identifying the target in each run.
Amplification Control (AC)	An artificially constructed DNA or RNA oligonucleotide containing target primer regions. It is added to an RT-PCR mastermix prior to aliquoting.	To verify that the RT-PCR has worked correctly in each individual reaction.

[a] Following ISO 22174:2024. Microbiology of the food chain – Polymerase chain reaction (PCR) for the detection and quantification of microorganisms – General requirements and definitions. This standard stipulates a list of requirements necessary for PCR-based methods for the detection and quantification of microorganisms in food samples. ISO 22174 defines a comprehensive list of controls which should be used in PCR-based analysis.

Controls for Sample Treatment

To verify that a sample treatment has worked correctly in each instance, an IPC is necessary. This is a non-target virus, as similar as possible to the target virus so that it effectively mimics the target's progress through sample treatment and extraction – if the IPC is successfully detected at the end of the method, the whole process including treatment, extraction, and detection can be considered as having worked correctly. The IPC can control for failure throughout the whole analysis; however, it is also best practice to an AC (see further on). Ideally, the IPC should be added immediately upon sample receipt in the analytical laboratory. During sample treatment, which in most cases is a multi-step process, losses of target virus can occur at each step. Using an IPC also facilitates determination of the efficiency of a sample treatment procedure (the extraction efficiency, i.e. how much target has been recovered). The IPC virus is added in a defined number, and at the end of the method the number of IPC virus detected is compared with the number which was added, and thus the percentage efficiency of extraction is determined.

An NPC is used to show that no laboratory contamination has occurred during the sample treatment. Ideally, verified uncontaminated samples of the foodstuff to be analysed should be used. As it might be difficult to acquire such verified target-free samples, however, the NPC is generally just sterile buffer. The NPC can control for contamination through the whole method; however, it is general practice to additionally control the nucleic acid extraction procedure by the use of a negative extraction control (NEC). Again, this is generally sterile buffer or water.

An NEC is used to control the nucleic acid extraction procedure and is generally sterile water in place of the sample extract; however, this control is not absolutely necessary if the batch under test includes an NPC.

Controls for the Assay

RT-PCR is an exquisitely powerful technique, but very sensitive to carryover contamination and to inhibition by substances present in the food and extracted with nucleic acids. Therefore, it is imperative to verify the correct amplification performance. To verify that the RT-PCR procedure has functioned correctly, an AC is essential. This is a target or a non-target oligonucleotide added to an RT-PCR. An AC signal must be observed; otherwise, the reaction can be stated to have failed (unless the presence of the target signal is observed). The AC exists in several variants: external or internal, and heterologous or homologous.

Where an external amplification control or EAC is used, two separate sample reactions are used; the EAC is added to one but not both, so that its amplification does not compete with that of the target in the first reaction. An EAC signal must be observed in each reaction to which it was added. If neither target nor EAC signal is observed, the test has failed; Figure 4.18 illustrates this.

There are several disadvantages to this approach: (1) it doubles the number of sample reactions and therefore the cost; (2) it does not give an unambiguous assurance that the first reaction has not failed; and (3) as the EAC is normally identical to the target and uses the same primers (and probe in real-time PCR), it has the potential for cross-contamination and hence to cause false-positive results.

In contrast, where an internal amplification control (IAC) is used, the IAC is amplified in the same reaction as the target. An IAC is added to the mastermix of reagents and therefore is a component of each reaction. An IAC signal must be observed when there is no target signal present; otherwise the reaction (and therefore the test) has failed. Figure 4.19 illustrates this.

An IAC can be homologous, where it uses the same RT-PCR primers as the target (but not the same probe), or heterologous where different primers (and probe in the case of real-time PCR) are used. An RT-PCR with a heterologous IAC is essentially multiplex, but it will not compete with the target for reagents to potentially hinder complete target amplification

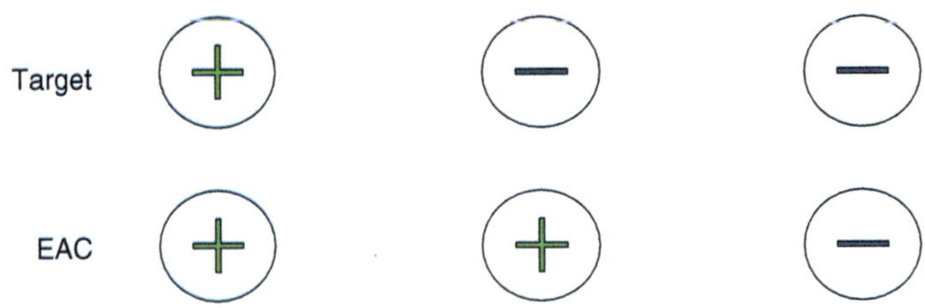

Figure 4.18 How to interpret results using the external amplification control.

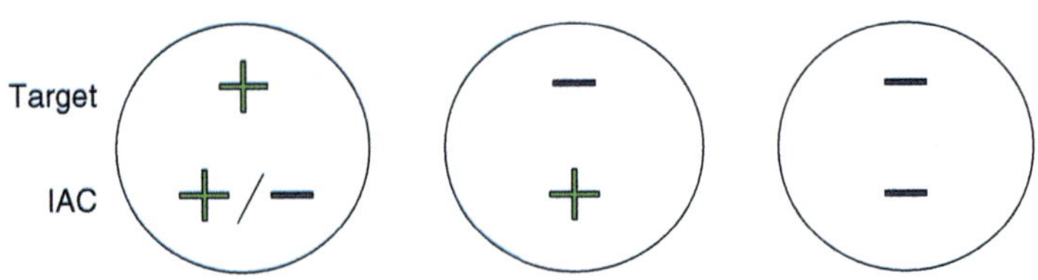

Figure 4.19 How to interpret results using the internal amplification control.

resulting in a reduced limit of detection (LOD). A homologous IAC will precisely mimic the amplification pattern of the target, and therefore fully control for inhibition and failure. A potential disadvantage of using a homologous IAC is that the IAC and target are competing for the same reaction components, and high concentrations of homologous IAC could prevent the amplification of low concentrations of target; however, this can be mitigated by careful optimisation of IAC concentration. The lowest concentration of IAC that will consistently produce a signal in the absence of target should be used. In samples containing large numbers of target, it will often be seen that no homologous IAC signal is observed, as the control has been outcompeted by the target. This is represented in the 'target-positive' in Figure 4.19.

ISO22174:2024 stipulates the use of a positive amplification control (PAC), and also commercially available RT-PCR assays have included the use of a PAC. This is a separate reaction, into which a known amount of nucleic acid from the target virus is included. Its purpose is to verify that the RT-PCR reaction functions correctly in all aspects, including the probe which is essential in real-time PCR. However, a PAC has a clear potential to cause false-positive results if errors are made in handling, which would result in PAC contamination of test reactions, and it is recommended that it is used with due caution.

Figure 4.20 shows the flow of controls in analysis of a food sample for viruses, detailing the stage of the procedure at which individual controls are incorporated.

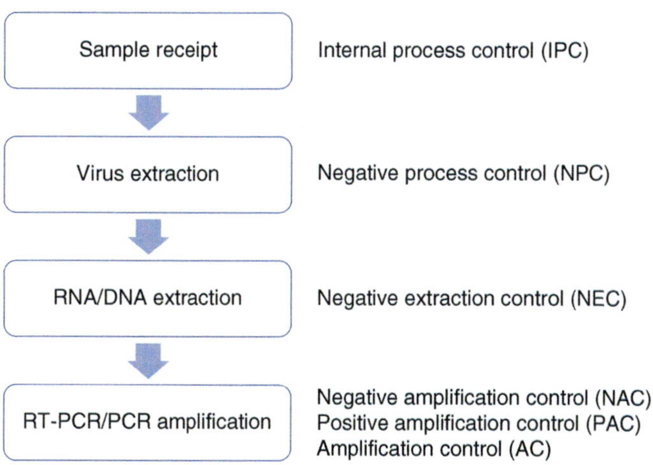

Figure 4.20 The flow of controls in a method to detect viruses in food.

Table 4.2 Interpretation of control results in the analysis of food samples for viruses.

	Result					
	Sample is positive			Sample is negative		
	Target	IPC[b]	AC	Target	IPC	AC
Sample	+	+	+/–[a]	–	+	+
NPC	–	+	+	–	+	+
NEC	–	+	+	–	+	+
NAC	–	N/A	+	–	N/A	+
PAC	+	N/A	+	+	N/A	+

Signal obtained (+), no signal (–). N/A = not applicable.
[a] When a homologous IAC is used, if a target signal is seen in the sample RT-PCR, but no IAC signal, and all other controls are correct, it should be accepted as a true positive, as high numbers of target may outcompete the IAC for amplification.
[b] In practice, if an IPC signal is not observed when a target signal is obtained, and all other controls show that the target signal is not due to contamination, then the sample can be recorded as a true positive, although it will not be possible to calculate the recovery efficiency.

Table 4.2 shows how to interpret the pattern of signals which must be observed when the sample is either positive or negative. Any other pattern of signals than the ones shown represents a failed analysis.

Correctly Interpreting the Readout from a Real-Time PCR Instrument

It is important to use the data produced by a real-time PCR instrument with care and attention. The experienced analyst will perform several checks before accepting the Ct values provided by the instrument. The cycle threshold can be set as default by the instrument. However, both the linear and logarithmic amplification plots must be checked to ensure that the threshold realistically represents the point at which amplification becomes exponential in all the reactions. The analyst should adjust the threshold if they consider it necessary. Checking each reaction plot will ensure that no spurious or anomalous amplification plots are accepted. Figures 4.21 and 4.22 show examples of each check including a spurious signal.

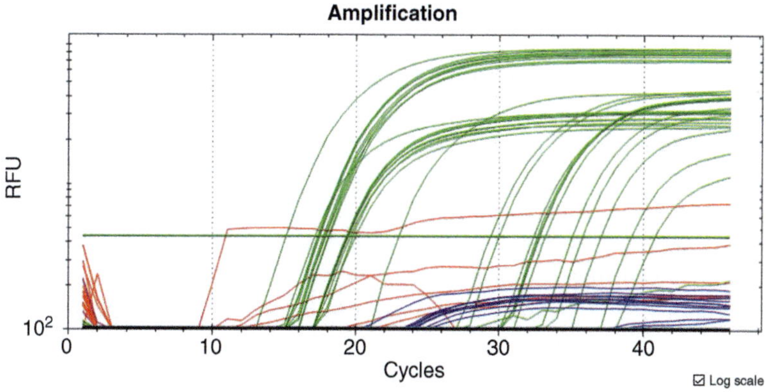

Figure 4.21 Logarithmic amplification plot of a sample set, illustrating typical target amplification curves. A non-specific (spurious) signal crosses the threshold cycle (Ct) at approximately Ct 11, indicative of potential background or artefactual amplification. *Source:* Martin D'Agostino.

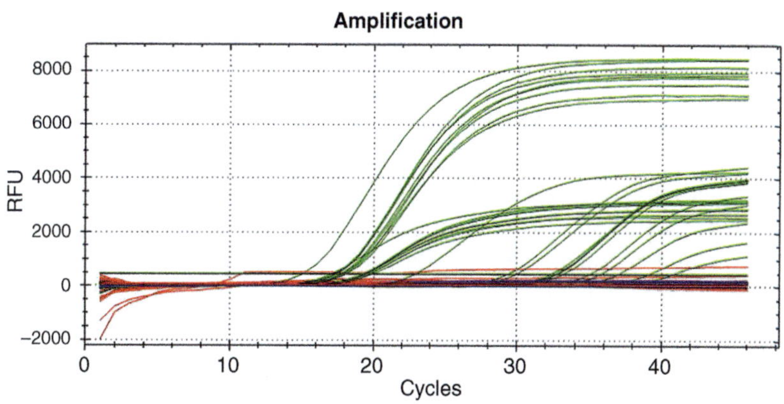

Figure 4.22 Linear amplification plot of the same sample set, displaying the fluorescence signal over successive cycles. The plot shows typical target amplification curves alongside a non-specific signal that becomes detectable at around cycle 11, which is more evident in the logarithmic scale representation in Figure 4.21. *Source:* Martin D'Agostino.

A Note Regarding High Ct Values

There is currently some debate on the interpretation of results from tests where 'high', e.g. Ct > 40, results are observed. If the amplification plots have been verified as correct, then high Ct values indicate that there were a small number of targets in the test reaction. It could be questioned whether such a low number of target viruses means that the foodstuff under test constitutes a risk to health. However, if a test result with a high Ct value happened to be derived from the amplification of even just one genome copy (which is theoretically possible), this does not necessarily equate to just one virus particle being present in the original sample tested: it only relates to the number of genome copies detected in one RT-PCR reaction. The quantity analysed in a single RT-PCR is typically 5% of the total nucleic acid extract (5 μl is taken from the 100-μl final extract of the sample and added to a well/tube containing mastermix). This theoretical one genome copy could thus equate to 20 genome copies in the 100-μl nucleic acid extract. As this extract represents the whole sample, it signifies that 20 virus particles were present in the sample, but it must also be borne in mind that the number of target viruses in any reaction is influenced by the efficiency of recovery. Therefore, if the efficiency of recovery was, e.g. 5%, our theoretical 20 virus particles actually equate to 400 in the sample. Even if only 5% of these 400 detected virus particles were infectious, this could signify that the tested foodstuff might pose a health risk. High Ct results should therefore not be dismissed out of hand.

The Issue of Infectivity

As stated in previous sections, RT-PCR has become the accepted technique for detection of viruses in food. However, it does not provide any information on the infectious nature of the viruses detected. An RT-PCR signal *per se* does not indicate whether the detected virus is infectious, as RNA can persist in inactivated viral particles if the capsid remains intact. This aspect is important for determination of any risk associated with consumption of virus-contaminated food. Infectivity is defined as the capacity of viruses to enter the host cell and exploit its resources to replicate and produce progeny infectious viral particles, which may lead to subsequent disease in the human host. Only cell culture–based methods offer a guarantee that the detected virus is infectious. However, suitable culture models for the most significant food viruses are not yet available for effective implementation, and it is unlikely that they can ever be used in routine food diagnostics, especially if high sample throughput is required. Therefore, considerable efforts have been made to adapt RT-PCR into an infectivity assay.

One approach is to exploit loss of capsid integrity in non-infectious particles. Degradation of the capsid by some forms of inactivation can make it

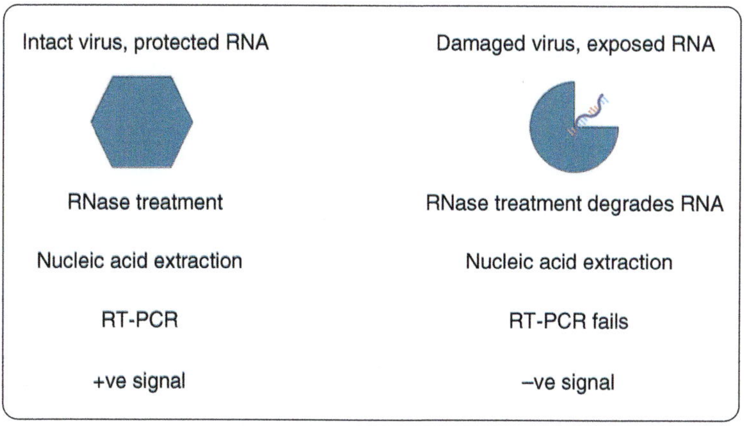

Figure 4.23 A schematic representation of the capsid integrity assay.

pervious to agents (e.g. nucleases) which can either eliminate the viral genomic material or prevent the nucleic acid from becoming amplifiable (e.g. through intercalation of dyes). The principle is that when damaged capsids are treated with these agents, then on the subsequent RT-PCR assay no positive signal will be produced. Conversely, only intact and therefore potentially infectious particles will be detected (Figure 4.23).

Another approach is to exploit possible fragmentation of the virus genome in non-infectious particles. Nucleic acid fragmentation by some forms of inactivation, e.g. ultraviolet irradiation, can prevent amplification (Figure 4.24). The general principle involves amplification of a long section of target nucleic acid sequence; if fragmentation has occurred, amplification will not be possible because the primer binding sites will be separated.

Another approach is binding of putatively infectious particles to virus-specific receptors (e.g. histo-blood antigen for norovirus) or substances that can mimic receptors (e.g. porcine mucin can bind norovirus particles). This is performed prior to RT-PCR, and subsequent target signals are interpreted as being derived from infectious particles (Figure 4.25).

Each of these approaches must be tested for each combination of virus, food matrix, and inactivation procedure. None of these techniques, however, can unambiguously selectively detect infectious virus. Inactivated virus might be intact or not substantially damaged, but still detectable; one can never be certain. The methods can only give an *indication* of possible infectivity, although even so, they still have value as being somewhat more informative than RT-PCR alone.

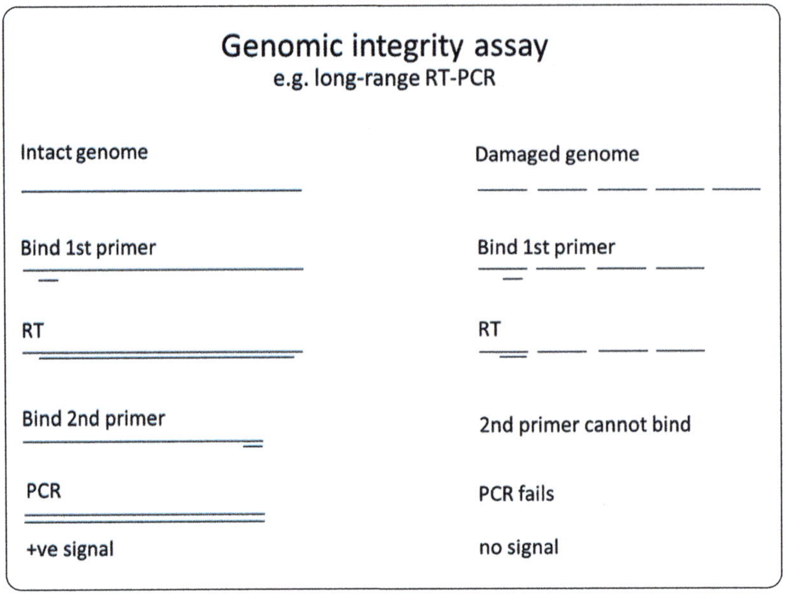

Figure 4.24 A schematic representation of the genomic integrity assay.

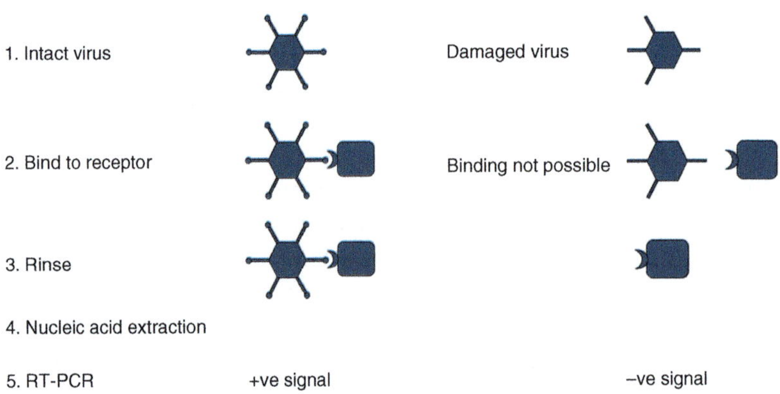

Figure 4.25 A schematic representation of the receptor-binding assay.

Virus infectivity assays are useful for studying the persistence of viruses in foods or in the environment and for evaluating the effect of disinfection procedures. However, it is debatable whether they are actually required in routine monitoring. Detection of viral genomic material by RT-PCR in a food sample means that the sample has at some stage been contaminated by that virus. For products such as berry fruit or leafy green vegetables, since virus contamination can only occur extrinsically, this means that some issues with agricultural practice and/or failure of hygienic practice have occurred somewhere along the supply chain. Controls to prevent or reduce the impact of such occurrences are discussed in Chapter 6.

Other Methods Utilising Nucleic Acid Amplification

Nowadays, there are other molecular methods which are available as alternatives to RT-PCR, such as nucleic acid sequence–based amplification technique (NASBA) or loop-mediated amplification (LAMP), which have the potential for virus detection in food. Although they offer speed and accuracy comparable to RT-PCR, their use in routine food testing requires further methodological development. Next-generation sequencing (NGS), which can analyse many genomic sequences simultaneously, has the theoretical potential to detect all viruses which may be present in a test sample of food; the realisation of this potential, however, will require very considerable refinements of existing sample preparation procedures.

Full References for Cited Standards

International Standards Organisation (2017). ISO 15216-1:2017. Microbiology of the food chain – horizontal method for determination of hepatitis A virus and norovirus using real-time RT-PCR – part 1: method for quantification. International Organization for Standardization, Geneva.

International Standards Organisation (2019). ISO 15216-2:2019. Microbiology of the food chain – horizontal method for determination of hepatitis A virus and norovirus using real-time RT-PCR – part 2: method for detection. International Organization for Standardization, Geneva.

International Standards Organisation (2024). ISO 22174:2024. Microbiology of the food chain – polymerase chain reaction (PCR) for the detection and quantification of microorganisms – general requirements and definitions. International Organization for Standardization, Geneva.

US Food and Drug Administration (2022). Bacteriological analytical manual chapter 26. Concentration, Extraction and Detection of Enteric Viruses from Food.

Further Reading

D'Agostino, M. et al. (2011). Nucleic acid amplification-based methods for detection of enteric viruses: definition of controls and interpretation of results. *Food and Environmental Virology* 3: 55–60.

Kim, T.Y. et al. (2023). A review of nucleic acid-based detection methods for foodborne viruses: sample pretreatment and detection techniques. *Food Research International* 174: 113502.

Liu, P. and Moore, M.D. (2020). A survey of analytical techniques for noroviruses. *Food* 9: 318.

Schmid, D. et al. (2004). Fast detection of Noroviruses using a real-time PCR assay and automated sample preparation. *BMC Infectious Diseases* 4: 15.

5

Prevalence and Survival

It is clear from the occurrence of outbreaks, product recalls, and notifications of contamination that viruses can be present in at-risk foods. There have also been some surveys in which viruses have been detected in several food items. Nonetheless, the prevalence of enteric viruses in the food supply is an incomplete picture. Effective methodology for the detection of viruses in foods has not been available for as long as methods for the detection of foodborne bacteria have been available, and large surveys for viruses are cost and labour intensive. However, it is essential to have comprehensive information on virus prevalence in order to be able to comprehend the full risk to public health and to be able to formulate effective procedures for prevention of virus contamination and elimination of viruses from contaminated foodstuffs.

The longer a virus can survive outside a host, the greater its chances of transmission. Outside a host, viruses can be considered as inert particles. They are not 'alive' in the strictest sense, and thus they do not die. They are either infectious or non-infectious. In this context, virus 'survival' signifies the persistence of infectious particles. Virus survival will be affected by various environmental conditions and factors such as heat, humidity, and pH. Enteric viruses are very resistant to desiccation, to ionising radiation, and to low temperatures (even to freezing), and this toughness further contributes to their success as pathogens. The robustness of enteric viruses allows them to remain infectious during the different situations and conditions that they can encounter while spreading from one host to another.

Prevalence in Foods and Food Production Settings

As seen in Chapter 2, enteric viruses cause a significant number of foodborne disease outbreaks. The World Health Organisation has estimated that 600 million cases of disease occur annually due to the consumption of contaminated foods, with norovirus being responsible for 124.8 million of these cases (~20% of total) and hepatitis A virus (HAV) responsible for 13.7 million (~2%). In the United States, it has been estimated that each year there are around 9.5 million episodes of domestically acquired (i.e. not from foreign travel) foodborne illness, around 5.5 million of which are caused by viruses, 99% of these being caused by norovirus. In the United Kingdom, the number of annual foodborne illness cases has been estimated as 2.4 million, of which 0.38 million can be attributed to norovirus. However, this is just the number of the reported cases for foodborne virus disease, which quite possibly are underestimated (see Chapter 2), and it will not precisely reflect the prevalence of enteric viruses in the food chain. Nonetheless, these data do show that enteric viral pathogens may be highly prevalent in at-risk foods. It is, however, difficult to say what the exact prevalence might be, or even what fluctuations in prevalence may occur year-to-year or season-to-season, as there are few programs of routine monitoring conducted nationally or internationally, and there is currently a lack of global harmonisation in sampling and detection approaches (see Chapter 4).

There have been surveys in some countries of at-risk foodstuffs for viruses. Most of these surveys have been conducted on a small-scale as part of research projects; some of them also examined environmental samples relevant to the food supply chain. The data obtained from these surveys have varied in regard to the food commodity tested and the geographic location of the study. When surveys of berry fruits and leafy green vegetables sold at retail have been conducted, the prevalence of norovirus has varied between 1% and around 50% (occasionally no contamination is found). Moreover, survey design is not harmonised, and in some surveys, data have been obtained from only very low numbers of samples. Norovirus has also been detected on dates and tomatoes. At the time of writing this book, there has only been a very limited number of studies examining the prevalence of HAV in fresh produce. When 858 samples of leafy green vegetables were examined in Italy in 2016, 2% were positive, and in Mexico 23% of 30 samples examined in 2015 contained HAV. A survey of leafy green vegetables in Egypt found HAV to be present in 31% of the samples tested. However, no HAV contamination was detected in a survey of 661 berry fruits conducted by the Canadian Food Inspection Agency in Canada in 2013. Since these surveys were not harmonised, i.e. they used different sample numbers and methodological approaches, it is not possible from the data they acquired to estimate the mean global prevalence of virus contamination in the food commodities examined.

Human infections caused by contaminated shellfish are reported worldwide, and the number of outbreaks of foodborne infections due to bivalve mollusc consumption is increasing. So far, the largest ever recorded outbreak of foodborne illness involved nearly 310,000 people. It occurred in China and was associated with the consumption of raw clams contaminated with HAV. The most common viruses monitored in bivalve molluscs are norovirus and HAV, although hepatitis E virus (HEV) has also been detected. The overall worldwide prevalence of enteric viruses in shellfish has been estimated at around 29% for norovirus. In a survey of 50 oyster production areas in the United Kingdom between 2009 and 2011, norovirus was detected in 76% of 844 samples, with at least one contaminated sample being detected from every site. Even if shellfish originate from approved growing areas or undergo depuration treatment before marketing, they still may contain enteric viruses with norovirus (NoVs) as the most frequently detected virus species. In marketed shellfish, a seasonal virus distribution pattern has been observed, with lower levels of NoV contamination present in mussels harvested during spring–summer compared to the autumn–winter season. HEV has been detected in pork products (liver, sausages) sold in retail in several countries including Canada, the Czech Republic, France, Italy, Senegal, Spain, Thailand, the United Kingdom, and the United States: an approximate prevalence of around 5% in commercially available pig livers was documented in these surveys. A nationwide study of pig liver samples taken at slaughterhouses in Spain found an overall prevalence of HEV of above 20%. In pig liver sausage and raw sausage from Germany, HEV detection rates have been reported between 20% and 22%. Particularly high detection rates (57.1–58.3%) have been described for a local liver sausage from France called Figatellu. HEV has also been detected in foods such as cow's milk, goat's milk, lettuce, raspberries, and seafood, but with lower detection rates.

Routine monitoring for foodborne viruses would allow a fuller picture of enteric pathogenic virus contamination to be acquired. However, monitoring is currently seldom performed routinely, particularly for fresh produce. Moreover, routine monitoring for specific foodborne viruses, e.g. norovirus and HAV, may not give a complete picture of the potential for foods to be contaminated with viruses. These viruses may be present as contaminants only very sporadically, or during outbreaks, and might be seldom detected even when food supply chains are vulnerable to contamination. It would be more effective to monitor for agents that would indicate that a route exists which norovirus and hepatitis A viruses could follow from source (human or animal) to points within the food supply chain and demonstrate therefore that vulnerabilities to virus contamination exist in that food supply chain. There are no suitable bacterial indicators for foodborne viruses, in the way that faecal coliforms can be used for foodborne bacterial contamination. The presence of other enteric viruses (route-tracking viruses) would

Table 5.1 Detection of a route-tracking virus in raspberry samples.

Raspberries	Number of samples tested	Norovirus detected	HAV detected	Human adenovirus detected
Fresh	60	0	0	1[a]
Frozen	95	0	0	3

[a] Porcine adenovirus was detected in four of the fresh berry samples, showing that a potential for contamination with HEV also existed.
Source: Data from Maunula et al. (2013).

clearly demonstrate that routes of contamination exist from the gastrointestinal tracts of humans or animals to the points in the food supply chain from where the samples are taken. Table 5.1 shows an example taken from a study in Europe where no norovirus or HAV was detected in any sample of raspberries, but several samples were positive for another enteric virus type.

Besides being useful as a route-tracking adenovirus in food of plant origin, human or animal adenoviruses are also considered as indicators of faecal contamination of shellfish growing waters, indicating a possible risk of NoV or zoonotic HEV occurrence. For instance, porcine adenoviruses have been used as indicators of food contamination with enterically transmitted zoonotic viruses of pig origin.

Monitoring for route-tracking viruses in food surveys would be very useful in defining the full potential for pathogenic virus prevalence. However, there is no single virus type that has been confirmed as being solely applicable as a route-tracking virus, and it may prove to be necessary to use a suite of virus types (see Chapter 7).

Natural Persistence

The number of outbreaks attributable to virus-contaminated food illustrates not only the prevalence of contamination, but also the potential for enteric viruses to persist for long periods in at-risk foodstuffs. There is a considerable body of information on enteric virus persistence in various environments, much of it gathered in early studies using cultivable enterovirus types such as coxsackievirus and poliovirus. In general, studies to determine the potential for persistence of viruses have been conducted using basic principles in common. Viruses in defined numbers are artificially introduced into a sample of water, soil, food, etc., and the sample stored under conditions (e.g. temperature, moisture) relevant to those experienced naturally, for selected time periods. Viruses are then extracted from the sample and infectious units and/or genome equivalents enumerated. The number of viruses remaining in the sample is then compared with the number which was introduced, and statistical procedures can be performed to calculate any rate of decline.

As mentioned previously, some of the main enteric viruses are not able, or are difficult, to be cultivated in the laboratory. Although there is a cell culture system which can be used to study HAV infectivity using a laboratory-adapted strain, there is still only a limited amount of information available on the persistence of this virus in foods or food production settings. Until very recently, there have been no available *in vitro* systems for determining HEV or norovirus infectivity, and most studies on the natural persistence of these agents have used quantitative polymerase chain reaction (PCR)-based methods to enumerate virus genomic units. Where a decline in numbers of genome equivalents is observed, a reliable inference is that a decline has also occurred (to the same extent or greater) in the number of infectious virus particles. However, no reliable inference on persistence of infectivity can be made when no decline in genome equivalents is observed, as infectivity can be lost but the genomic material can remain detectable.

Survival on Food Contact Surfaces

Enteric viruses can retain infectivity on food contact surfaces (e.g. such as those used for food preparation) over prolonged periods. Humidity may increase virus survival, depending on the temperature: the combination of low (4–5°C) temperatures and high humidity is ideal for persistence of infectivity. In moist environments, HAV can persist on surfaces such as stainless steel for at least 4 days at 20°C. It is possible that norovirus can persist for at least 7 days and perhaps for considerably longer on food preparation surfaces made from materials including ceramic, Formica™, PVC, and stainless steel, especially at high humidities and low temperatures such as those used in refrigerated storage conditions. Some studies have indicated that norovirus GII may be more stable on surfaces than norovirus GI, although the reasons for this are unknown. HAV is resistant to desiccation, and infectious virus can be detected on dry stainless steel even after 28 days at room temperature. Resistance to desiccation appears to be significant in enabling enteric viruses to survive on surfaces and may account in some part for the seasonality of infections which has been observed for some virus types. The seasonality of infections with norovirus and rotavirus may be due in part to the low relative humidity indoors during the cold periods when houses and buildings are heated, encouraging the persistence of infectious viruses on surfaces.

HAV can survive in an infectious state for at least 2 hours on hands. The persistence of viruses on environmental surfaces and their subsequent ability to transfer to hands may be important factors in the spread of foodborne viruses and explain in some part the unidentified mode of transmission in many investigated outbreaks.

Survival in Water

Several studies have established that enteric viruses can survive for weeks in various types of waters. The major influencing factors appear to be temperature, ultraviolet (UV) light, the presence of particulate material, and microbial predation. For example, several weeks can be necessary for a 99% reduction of infectious HAV in several types of water (e.g. ground, river), with the presence of microorganisms having a negative effect and the presence of particulate material, such as clay, enhancing survival. If mineral water is contaminated by HAV and stored at 4°C, infectious virus particles might still be present after 1 year. Norovirus has been shown to have the potential to persist in an infectious state in contaminated groundwater for at least 8 weeks. Norovirus GI may be more persistent in water than is norovirus GII. Adenovirus, coxsackievirus B1, and poliovirus have been observed to remain infectious for at least 1 year in artificially contaminated groundwater.

Survival on Crops

The use of sewage-contaminated water for irrigation is considered a potential route of virus contamination of produce. It has been demonstrated that on vegetables such as celery, lettuce, radishes, spinach, and tomatoes, poliovirus introduced via spraying with contaminated water could persist for up to at least 5 weeks until the crops were picked, although 90–99% losses could occur during the first week, and rainfall could remove the virus from plant surfaces. Solar radiation has a negative influence on the persistence of viruses on irrigated crops. Modelling of pre-harvest conditions encountered in climates such as those of the United States and Central America has indicated that HAV might survive on crops such as bell peppers, cantaloupe, and lettuce for several months. It is likely that under the growing conditions encountered in colder climates, e.g. Northern Europe, the survival of enteric viruses on crops could be at least as robust.

Survival in Food

Studies have shown that enteric viruses can survive on fresh berries such as raspberries and strawberries for several days under the conditions (refrigeration or room temperature) commonly used during household storage. Furthermore, the evidence from several outbreaks of hepatitis A and norovirus gastroenteritis strongly indicates that viruses can survive for several months in frozen berries (see Chapter 3). The acidity of the fruit will have little effect as enteric viruses are tolerant of acidic conditions even as low as pH 1. Freeze-drying, such as is commonly performed to dry berry fruit for subsequent use as food additives, can reduce the number of infectious virus on berry fruits, but if the viruses are present in high numbers, some could still remain on the berries after the treatment.

HAV can survive on fresh vegetables such as carrots, green onions, fennel, lettuce, and spinach for at least 1 week at refrigeration temperatures, indicating that the virus can persist under normal refrigeration conditions during the time that will elapse between purchase and consumption. It has also been shown that, if contaminated foods are stored at room temperature, around 10% of infectious virus can survive. It is possible that refrigeration temperatures can enhance the survival of norovirus on lettuce as compared with storage at room temperature, with norovirus GII being more persistent than GI. Modified atmospheres (e.g. high CO_2), commonly used in packaging fresh-cut leafy vegetables to supress the growth of spoilage bacteria, might actually enhance virus survival by suppressing inhibitory microflora or enzymes.

The numerous outbreaks which have occurred demonstrate that HAV and norovirus can survive in bivalve shellfish such as oysters, mussels, and clams from contamination to the point of consumption. Infectious HAV can persist for at least 3 weeks in shellfish from the point of contamination, while the viral RNA could be detected even after 6 weeks. Studies with other enteric viruses such as poliovirus have shown that viruses can survive in oysters and mussels stored at 4–5°C for at least 28 days – the longest time normally employed between harvesting and consumption. The ability of enteric viruses to survive in chilled and frozen shellfish represents a potential health risk and is significant both to the shellfish industry and public health agencies. The persistence of enteric viruses in bivalve shellfish appears to be associated with the ability of the viruses to become sequestered within shellfish tissue, with the duration of persistence related to the stability of the virus within the acidic environment of the tissues. It has also been shown that HAV and potentially norovirus could survive in marinated

Table 5.2 Survival of HAV and norovirus on foodstuffs and food production environments (at least 1% virus remaining).

Food type	Hepatitis A virus	Norovirus
Berries (stored frozen)	Months[a,c]	Weeks to months[c]
Berries (stored refrigerated)	NI	Weeks[b]
Berries (stored at ambient)	NI	Days[b]
Deli meats	Hours[c]	Days to weeks[b]
Handlers' hands	Hours[a]	Hours[b]
Lettuce (stored refrigerated)	Days[a]	Weeks[b]
Lettuce (stored at ambient)	Days[a]	Days[b]
Preparation surfaces (e.g. formica, stainless steel)	Days[a]	Days to weeks[b]
Raw shellfish	Days to weeks[a,c]	Days to weeks[b,c]
Salad onions (stored refrigerated)	Days[a,c]	Days to weeks[c]
Salad onions (stored at ambient)	Days[a]	NI
Water	Days to weeks[a]	Weeks[b]

NI, no information.
[a] Information derived from studies on persistence of infectious virus.
[b] Information derived from studies on persistence of detectable norovirus RNA; infectivity may not persist as long.
[c] Derived from observations of outbreak details.

mussels, thus constituting a potential health risk if shellfish are contaminated prior to this form of processing.

It is currently difficult to determine the survival characteristics of HEV. However, studies have indicated that storage of pork liver under refrigeration for several weeks did not result in a considerable decline in the number of detectable viral genomes, indicating that substantial numbers of HEV particles remained intact and thus potentially infectious.

In the past, milk has been implicated as a vehicle for the transmission of poliovirus, due to contamination by infected persons. In pasteurised milk, yogurt, and cottage cheese, poliovirus can survive for weeks at 4°C. Poliovirus inoculated into cheese could persist over 7 months with less than 90% decline; however, the cheese-making process could be sufficient to control any viral contamination which might be present in the raw milk.

Studies have shown that HAV could survive on low-moisture foods such as biscuits stored at ambient temperature for many days and that norovirus may be able to persist in food products such as cooked turkey and other processed foods such as minced meat and potato salad for at least 7 days at refrigeration temperatures.

As well as the experimental data from survival studies, features of some outbreak descriptions reveal the potential for robust survival of enteric viruses in foods. For example, the strawberries implicated in the outbreak of HAV in the United States in 1997 (see Chapter 3) had been harvested in the Spring of the previous year and stored frozen for several months prior to consumption in February 1997. Freezing appears to have little or no effect on enteric virus infectivity or integrity, and experimental and anecdotal evidence indicates that norovirus and HAV can remain intact even after several freeze–thaw cycles.

Table 5.2 gives basic estimates of how long HAV and norovirus could persist in various foods and food production settings. It is clear that foodborne viruses can survive from a contamination event, even if this occurs at primary production, to the point of consumption, thus causing the contaminated food to constitute a hazard to the consumer.

Sources of Quoted Data

Epidemiology: World Health Organization (2015). *Estimates of the Global Burden of Foodborne Diseases: Foodborne Disease Burden Epidemiology Reference Group 2007–2015*. Geneva, Switzerland: World Health Organization. ISBN: 9789241565165. FAO-WHO (2024). Microbiological risk assessment of viruses in foods, Part 1: Food attribution, analytical methods and indicators. Microbiological Risk Assessment Series number 49; Scallan, E. et al. (2011). Foodborne illness acquired in the United States – major pathogens. *Emerging Infectious Diseases* 17: 7–15. Daniel, N. et al. (2020). *The Burden of Foodborne Disease in the United Kingdom in 2018*. Food Standards Agency. ISBN: 978-1-78256-530-5.

Surveys of viruses in produce: Terio, V. et al. (2017). Occurrence of hepatitis A and E and norovirus GI and GII in ready-to-eat vegetables in Italy. *International Journal of Food Microbiology* 249: 61–65. Parada-Fabián, J.C. et al. (2016). Identification of enteric viruses in foods from Mexico City. *Food and Environmental Virology* 8: 215–220. Elmahdy, E.M. et al. (2022). Detection of norovirus and hepatitis A virus in strawberry and green leafy vegetables by Using RT-qPCR in Egypt. *Food and Environmental Virology* 14: 178–189. Steele, M. (2022). Norovirus GI and GII and hepatitis A virus in berries and pomegranate arils in Canada. *International Journal of Food Microbiology* 379: 109840.

Outbreak of hepatitis in China associated with clams: Halliday, M.L. et al. (1991). An epidemic of hepatitis A attributable to the ingestion of raw clams in Shanghai, China. *Journal of Infectious Diseases* 164: 852–859.

Surveys of viruses in shellfish: Li, Y. et al. (2023). A systematic review and meta-analysis indicates a substantial burden of human noroviruses in shellfish worldwide, with GII.4 and GII.2 being the predominant genotypes. *Food Microbiology* 109: 104140. Lowther, J.A. et al. (2012). Two-year systematic study to assess norovirus contamination in oysters from commercial harvesting areas in the United Kingdom. *Applied and Environmental Microbiology* 78: 5812–5817.

Surveys of HEV in pork and other meats: Harrison, L. et al. (2021). Presence of hepatitis E virus in commercially available pork products. *International Journal of Food Microbiology* 339: 109033. Pavlova, A. et al. (2025). Hepatitis E virus in the role of an emerging food-borne pathogen. *Microorganisms* 13: 885. Mykytczuk, O. et al. (2017). Prevalence and molecular characterization of the hepatitis E virus in retail pork products marketed in Canada. *Food and Environmental Virology* 9: 208–218. Szabo, K. et al. (2015). Detection of hepatitis E virus RNA in raw sausages and liver sausages from retail in Germany using an optimized method. *International Journal of Food Microbiology* 215: 149–156. Tene, S.D. et al. (2024). Detection of hepatitis E virus (HEV) in pork sold in Saint-Louis, the North of Senegal. *Life (Basel)* 14: 512. Thippornchai, N. et al. (2022). Survey of hepatitis E virus in pork products and pig stools in Nakhon Pathom Province, Thailand. *Veterinary Medicine and Science* 8: 1975–1981.

Monitoring for route-tracking viruses: Kokkinos, P. et al. (2012). Harmonised investigation of occurrence of human enteric viruses in the leafy green vegetables supply chain in three European countries. *Food and Environmental Virology* 4: 179–191; Maunula, L. et al. (2013). Tracing enteric viruses in the European berry fruit supply chain. *International Journal of Food Microbiology* 167: 177–185.

Natural persistence: Rzeżutka, A. and Cook, N. (2004). Survival of human enteric viruses in the environment and food. *FEMS Microbiology Reviews* 28: 441–453.

Survival on food contact surfaces: Bae, S.C. et al. (2014). Survival of hepatitis A virus on various food-contact surfaces during 28 days of storage at room temperature. *Food Research International* 64: 849–854. D'Souza, D.H. et al. (2006). Persistence of caliciviruses on environmental surfaces and their transfer to food. *International Journal of Food Microbiology* 108: 84–91. Liu, P. et al. (2009). Persistence of human noroviruses on food preparation surfaces and human hands. *Food and Environmental Virology* 1: 141–147. Mbithi, J.N. et al. (1991). Effect of relative humidity and air temperature on persistence of hepatitis A virus on environmental surfaces. *Applied and Environmental Microbiology* 57: 1394–1399. Terpstra, F.G. et al. (2007). Resistance of surface- dried virus to common disinfection procedures. *Journal of Hospital Infection* 66: 332–338.

Survival in water: Biziagos, E. et al. (1988). Long-term persistence of hepatitis A virus and poliovirus type 1 in mineral water. *Applied and Environmental Microbiology* 54: 2705–2710. Bosch, A. (1995). The persistence of enteric viruses in the water environment. *Microbiologia* 11: 393–396. Trudel-Ferland, M. et al. (2021). Persistence of hepatitis A virus RNA in water, on non-porous surfaces, and on blueberries. *Frontiers in Microbiology* 12: 618352; Sobsey, M.D. et al. (1986). Persistence and transport of hepatitis A virus in soils, groundwater and wastewater. *Water Science and Technology* 10: 97–106.

Survival on crops: Bagdasaryan, G.A. (1964). Survival of viruses of the enterovirus group (poliomyelitis, echo, coxsackie) in soil and on vegetables. *Journal of Hygiene, Epidemiology, Microbiology, and Immunology* 8: 497–505. Tierney, J.T. et al. (1977). Persistence of poliovirus 1 in soil and on vegetables grown in soil previously flooded with inoculated sewage sludge or effluent. *Applied and Environmental Microbiology* 33: 109–113.

Survival in food: Bidawid, S. et al. (2001). Persistence of hepatitis A virus on modified atmosphere-packaged (MAP) lettuce. *Food Microbiology* 18: 95–102. Butot, S. et al. (2008). Effects of sanitation, freezing and frozen storage on enteric viruses in berries and herbs. *International Journal of Food Microbiology* 126: 30–35. Cliver, D.O. (1973). Cheddar cheese as a vehicle for viruses. *Journal of Dairy Science* 6: 1329–1331. Croci, L. et al. (2002). The survival of hepatitis A virus in fresh produce. *International Journal of Food Microbiology* 73: 29–34. Di Girolamo, R. et al. (1970). Survival of virus in chilled, frozen, and processed oysters. *Applied Microbiology* 20: 58–63. Hewitt, J. and Greening, G. (2004). Survival and persistence of norovirus, hepatitis A virus, and feline calicivirus in marinated mussels. *Journal of Food Protection* 67: 1743–1750. Kurdziel, A.S. et al. (2001). Survival of poliovirus on soft fruit and salad vegetables. *Journal of Food Protection* 64: 706–769. Lorusso, P. et al. (2024). Evaluation of hepatitis E virus RNA persistence in experimentally contaminated cured pork liver sausages. *Italian Journal of Food Safety* 13: 12286. Monini, M. et al. (2024). Persistence of hepatitis E virus (HEV) subtypes 3c and 3e: Long-term cold storage and heat treatments. *Food Microbiology* 121: 104529. Tiron, S.V. (1992). Field and laboratory studies related to the persistence, survival and inactivation of enteroviruses in some foods. *Proceedings of the Third World Congress on Food-borne Infections and Intoxications*, Vol. 1, pp. 298–303.

Further Reading

Cook, N. et al. (2016). Persistence and elimination of human norovirus in food and on food contact surfaces: a critical review. *Journal of Food Protection* 79: 1273–1294.

European Commission. Rapid alert system for food and feed (RASFF). (https://food.ec.europa.eu/food-safety/rasff_en)

European Food Safety Authority (EFSA) (2019). *Scientific report on analysis of the European baseline survey of norovirus in oysters. EFSA Journal* 17 (7) 5762, 86 pp. https://doi.org/10.2903/j.efsa.2019.5762.

Morales-Rayas, R. and Jean, J. (2012). Overview of viral persistence on food surfaces. *Current Topics in Virology* 9: 1–17.

Trudel-Ferland, M. (2021). Persistence of hepatitis A virus RNA in water, on non-porous surfaces, and on blueberries. *Frontiers in Microbiology*, 12, 618352. https://doi.org/10.3389/fmicb.2021.618352

6

Control of Contamination

Procedures to eliminate viruses, or their infectivity, can be applied if a foodstuff is suspected of being exposed to virus contamination. Various procedures can be implemented, although not all are effective or are appropriate for specific food types. Furthermore, some food components can have a protective effect on virus infectivity. Ideally, however, viruses should be prevented from entering the food supply chain in the first instance. Prevention is preferable to trying to eliminate contaminating viruses from food, particularly as elimination is not always feasible, especially with foods intended to be eaten raw or after minimal processing. Preventive practices can and should be applied at all stages of the food supply chain – production, processing, retail, catering, and in the home. A thorough assessment of the risk of foodborne virus contamination can assist in identifying the best form of prevention or, if necessary, identifying an optimal elimination process for use at specific stages of the food supply chain.

Elimination of Viruses from Food

The commonly accepted aim of disinfection procedures used in the food industry is to reduce pathogen load by at least 4 log (99.99% reduction). The principal measures employed to eliminate unwanted microorganisms in food supply chain settings are based on chemical disinfection and other physical procedures such as ultraviolet irradiation. Among chemical disinfection procedures, the use of chlorine is most common. Heat is the most commonly applied physical procedure, although alternative non-thermal techniques, particularly high hydrostatic pressure, are becoming adopted.

Evaluation of the effectiveness of a disinfection procedure on viruses is performed in a similar way to evaluation of virus survival. Viruses in defined numbers are artificially introduced into a sample of foodstuff or, for instance, onto a surface, and then the disinfection procedure is applied. The number of infectious virus or virus particles remaining in the sample is then compared with the original number. The fewer viruses that remain, the more effective the disinfection procedure. However, as mentioned in Chapter 4, effective methods for evaluating HEV and norovirus infectivity are not currently widely available, and therefore data on the effect of elimination procedures are not extensive for these virus types. A body of experimental work has been performed using surrogate viruses (e.g. feline calicivirus and murine norovirus) as models for human norovirus. These surrogates need to be directly evaluated against infectious human norovirus using a cell culture-based system to be able to say definitively that they mimic human norovirus in regard to persistence under various environmental conditions or processing treatments.

Chemical Disinfection

The efficacy of disinfection agents against viruses is influenced by the contact time, concentration of disinfection agent, and the type of virus involved. Specific environmental factors such as temperature, humidity, and pH can also affect efficacy. The characteristics of viruses influence the range of inactivation by disinfection. Enveloped viruses contain a lipid envelope that is required for infection, and therefore interfering with the envelope should reduce virus infectivity. However, it is much more challenging to inactivate the major enteric viruses, which are non-enveloped viruses, and several commonly available disinfectants are not able to sufficiently reduce infectivity. Moreover, organic material or soil which could be found associated with food or in food-related environments can interfere with chemical disinfection. Furthermore, and more importantly, chemical disinfection of ready-to-eat or raw foods such as soft berries may not be appropriate, as their appearance, texture, and taste could be adversely affected.

Currently, the two main types of disinfectants which are most commonly used in the food industry are chlorine-based products and quaternary ammonium compounds (QACs). Chlorine can be a suitable disinfectant for foodborne viruses in some applications. Treatment of water with 10 ppm (parts per million) chlorine for 30 minutes prevented norovirus infection of volunteers. At 188 ppm chlorine, more than 3 logs of norovirus were removed from ceramic tile and stainless steel. Sodium hypochlorite at a concentration of 3% removed at least 4 logs of norovirus from stainless steel samples after 10 minutes contact. HAV required contact with 2500–5000 ppm sodium hypochlorite for 5 minutes to decline by 4 logs infectivity on stainless steel. On foods, however, the limited number of studies which have been performed has shown that chlorine-based disinfectants have a very limited effect. Free chlorine at 200 ppm in wash water reduced HAV in raspberry and strawberry samples by less than 2 logs. On HAV-contaminated iceberg lettuce, strawberries, and tomatoes, immersion for 5 minutes in water containing 200 ppm chlorine at 5 minutes was necessary to produce a 2.6-log infectivity reduction on berries and a 4-log infectivity reduction on lettuce. Using the same chlorine concentration, up to 3.4 logs of norovirus was eliminated from several berry types and fresh herbs. However, these chlorine concentrations are higher than are normally industrially employed for treatment of fresh produce.

QACs have been shown to be effective on surfaces against human adenovirus and coxsackie virus, but they have little effect on norovirus and poliovirus. Peracetic acid, another common surface disinfectant, has also been shown to be effective against infective HAV, but not against norovirus as determined by reverse transcription polymerase chain reaction (RT-PCR). Even at concentrations of 70% or higher, ethanol is not fully effective against all norovirus strains and HAV; consequently, purely alcohol-based hand-sanitisers cannot be relied on to ensure good hand hygiene.

Plant materials such as seeds and essential oils contain substances which have antimicrobial activity, e.g. flavonoids, phenolics, carotenoids, terpenoids, alkaloids, and the inexpensive availability of extracts of these materials could make them a promising alternative as a food disinfectant. Some plant extracts have some effectiveness against infectious HAV on leafy vegetables, inactivating up to ~3-log virus; however, little or no effect has been seen in studies against several other enteric viruses.

Table 6.1 gives a basic overview of the effectiveness of various disinfectants commonly used in the food industry against norovirus and hepatitis A virus (HAV). There are very scarce data for chemical disinfection of hepatitis E virus (HEV).

Physical Disinfection

Heat

For foods, the traditional procedure for microbial control has been the use of high temperature. This treatment is very efficient for pathogen reduction; however, although it is commonly applied to foods of animal origin, it is seldom used for ready-to-eat foods as it can drastically affect their sensorial and nutritional properties. In pure suspension, HAV is relatively resistant to high temperatures, e.g. those applied during pasteurisation. Few studies have been conducted on the effect of heat on HAV in foods; the available data indicate that some food components, e.g. fats and sugars, can increase the virus' resistance to higher temperatures. Commercial heat-treatment of shellfish, where the internal temperature is raised to 90°C for

Table 6.1 Generalised efficacy of commonly used chemical disinfectants against norovirus and HAV.

Disinfectant	Virus	Matrix	Efficacy[a]
Chlorine-based	Norovirus	Surfaces	High
		Foods	High
	HAV	Surfaces	High
		Foods	Low to medium
QACs	Norovirus	Surfaces	Low
		Foods	n.d.
	HAV	Surfaces	Low
		Foods	n.d.
Peracetic acid	Norovirus	Surfaces	Low
		Foods	Low
	HAV	Surfaces	Medium
		Foods	Low
Ethanol	Norovirus	Surfaces	Low to medium
		Foods	n.d.
	HAV	Surfaces	Low
		Foods	n.d.

HAV, hepatitis A virus; n.d., no data; QAC, quaternary ammonium compound.
[a] Efficacy rates: Low = log reduction <1.0; Medium = 1.0–3.0; High = >4.0.

90 seconds, is, however, effective for elimination of infectious HAV in these products. Heating at 85°C for 30 seconds can reduce infectious HAV in dairy products by 6 logs. Heating at 56°C for 1 hour was ineffective at eliminating HEV infectivity in pig liver tissue, and it may be necessary to cook liver at 71°C for at least 5 minutes to inactivate this virus. NoV may be resistant to temperatures below 60°C, but temperatures of 72°C and above can reduce NoV genome equivalents by at least 3 logs. Research into the effect of heat on enteric viruses is ongoing; however, currently there have not been sufficient studies performed using at-risk foods such as berry fruits or their products to precisely determine the effect of the procedures used in the food industry, and harmonisation of the available data is challenging. Table 6.2 gives a basic overview of the effect of heat on the major foodborne viruses. Please note that the generalised effectiveness of heat on norovirus is based on the detection of genetic material and not on the evaluation of infectivity.

Table 6.2 Generalised efficacy of heat against the main foodborne viruses.

Virus	Treatment[a]	Efficacy[b]
Norovirus	Pasteurisation	Low to medium
	High temperature	Medium to high
	Cooking	Medium to high
HAV	Pasteurisation	Low to medium
	High temperature	High
	Cooking	High
HEV	Pasteurisation	Medium
	High temperature	Medium to high
	Cooking	High

[a] Pasteurisation: treatment with a temperature/time equivalent to 75°C for 15 seconds; High temperatures: temperatures above 90°C; Cooking: culinary procedures such as frying, baking, boiling, etc.
[b] Efficacy rates: Low = log reduction <1.0; Medium = 1.0–3.0; High = >3.0.

Non-Thermal Treatments

As well as chemical disinfection and heating, other methods to reduce viral loads have been studied. Efforts have been made to find ways of using treatments which keep the integrity of the foods in their original state or as close as possible to it. This is especially beneficial for delicate foods such as soft berries or fruit juices.

Gamma Irradiation

A form of ionising radiation, gamma irradiation has been noted for its ability to degrade major capsid proteins, leading to significant virus reductions. The maximum irradiation dose which can be used for processing foods, according to the Food and Agriculture Organisation, is 10 kGy. However, this level may not be fully effective against foodborne viruses, and in many countries only lesser doses are permissible.

High Hydrostatic Pressure (HHP)

High hydrostatic pressure (HHP) is a non-thermal process that is increasingly used in the food industry to reduce the microbial load in some foodstuffs while preserving their sensorial properties. It is effective against foodborne bacteria, although viruses can display greater resistance. Treatments above 350 MPa for 5 minutes could reduce HAV infectivity in fruit puree and green onion slices. Pressures higher than 400 MPa can reduce HAV infectivity in shellfish by at least 3 logs. Treating oysters at 600 MPa for 5 minutes can eliminate norovirus infectivity, as demonstrated in a study using human volunteers. HHP can reduce infectious HEV in meat products by at least 2 logs; however, the reduction is low and further studies are necessary to produce data to allow the food industry to consider implementation of this approach. HHP is completely ineffective against poliovirus.

Ultraviolet Light (UV)

Ultraviolet (UV) light is widely used to disinfect surfaces in the food industry. It works mainly by causing damage to microbial genomic nucleic acids. In RNA viruses, and at high levels, it can also disrupt the capsid. Irradiation at 40 mW s cm^2 and above can eliminate over 4 logs infectious HAV on the surfaces of leafy green vegetables. It has been demonstrated to be very effective against HAV on surfaces. Its effectiveness against norovirus and HEV has not been fully evaluated. Its use for foodborne virus control is limited, as it will not inactivate viruses shielded by, e.g. textures or cracks in the surfaces of fresh produce items (e.g. raspberries) or internalised in meat products.

Depuration

Depuration is a purification process used by the shellfish industry to remove microbiological contamination. Harvested shellfish are placed into tanks containing fresh water to allow them to eliminate contaminants from their filters (Figure 6.1). Often this is done in relays, where the water is changed

Figure 6.1 Depuration tanks containing oysters. *Source*: Mark Stebnicki / Pexels.

after a certain period. The parameters for depuration have been set to mediate elimination of bacteria; however, studies have shown that these parameters may not be fully effective for virus removal. For example, it has been shown that 4-week depuration does not completely remove norovirus from oysters. Even if the shellfish are placed in tanks and exposed to UV light or ozone, the degree of elimination may not be sufficient. It has been recommended that extensive relaying may need to be performed to assure absence of viral pathogens from shellfish.

Prevention of Contamination

As seen in the previous section, disinfection of virus-contaminated foods may not be completely effective in all cases, and with extrinsic contamination at least, it may be a case of 'too little and too late'. Ideally, the primary focus of food microbiological hygiene should be on preventing pathogens from entering the food supply chain in the first instance. With intrinsic viral contamination, this is very challenging, but with extrinsic contamination it should be possible to achieve through the strict application of preventative procedures during food production.

Procedural Controls for Prevention of Contamination of Fresh and Ready-to-Eat Produce

At primary production of berry fruit and leafy vegetables, we can be certain that the main routes of HAV and norovirus contamination are handlers' hands and/or irrigation water and that the main route of HEV contamination will be contact of crops with pig excrement either applied as manure or via contaminated irrigation water. Therefore, control of virus contamination should be focussed on those critical points during primary production.

A range of preventative procedures are specified by prerequisite programs such as those based on Good Agricultural Practice (GAP), Good Hygienic Practice (GHP), and Good Manufacturing Practice (GMP) and listed in relevant guidelines such as those produced by the Codex Alimentarius Commission. Compliance with these recommendations and guidelines should greatly reduce the potential for virus contamination. The Food and Agricultural Organisation describes GAPs as 'practices that address environmental, economic and social sustainability for on-farm processes, and result in safe and quality food and non-food agricultural products'. However, no GAP guideline specifically mentions viruses. In 2012, the Codex

Alimentarius Commission produced the document 'Guidelines on the Application of General Principles of Food Hygiene to the Control of Viruses in Food', based on the existing Codex general principles of food hygiene. The Codex Guidelines emphasise prevention of contamination through various measures, including:

- Use of clean water for irrigation and washing
- Education of food handlers about microbial safety guidelines and hygiene rules
- Provision of adequate toilet facilities for agricultural workers
- Limiting access of young children to harvesting settings

All enteric viruses can contaminate water, e.g. through sewage and run-off from pig farms, and can be transmitted by water. Therefore, ensuring the sanitary quality of water is necessary to avoid the spread of virus contamination along the food supply chain. It is an especially critical point during the primary production of ready-to-eat foods. Several surveys have noted the presence of enteric viruses in waters, including waters used for irrigation of crops. Consequently, the design of the irrigation systems for produce must consider the avoidance of water which has the potential for contamination with human and animal faeces. In addition to that, during the processing of ready-to-eat produce, the water used in the washing steps must possess guaranteed sanitary quality, if necessary through the addition of chemical disinfectants (e.g. chlorination or ozonisation) or the use of physical treatment (e.g. UV).

All workers employed in handling food, at every stage of the food supply chain, should have training in microbiological food hygiene, and this should include education about the hazards of foodborne viruses. The essential aspect that must be emphasised in a training curriculum is personal hygiene, strongly highlighting the absolute necessity for proper hand washing. Effective hand washing has been demonstrated to reduce enteric viruses on hands by up to 100-fold. Food handlers' hands must be kept clean, with special attention paid to washing after using toilet facilities. Given the possibility that shedding of norovirus can occur asymptomatically and pre-symptomatically, food handlers must never assume that they cannot contaminate food if they do not feel unwell.

To guarantee that properly trained workers can implement thorough personal hygiene, food business operators must provide good toilet facilities (Figure 6.2). It can be noticeable that in primary production, good toilet facilities are not always provided. This is a major failing and increases the potential for contamination of the food with a range of microbial

Figure 6.2 An example of a good toilet facility. *Source*: Generated with AI using OpenAI.

pathogens, not just enteric viruses. Ensuring that workers have appropriate toilet facilities must be a priority for inclusion for primary producers to ensure compliance with guidelines and pre-requisite programs.

As stated in many food safety guidelines, persons not involved in food handling should not be present during food production from harvesting to processing. This especially applies to young children, who could in some instances (e.g. when living in highly endemic regions) be asymptomatically infected, e.g. with HAV (see Chapter 2). Food handlers should also be aware of the risks posed by contact with infected individuals at home or other settings.

The 2012 Codex Foodborne Virus Guidelines are focussed on HAV and norovirus and do not refer to HEV; at the time of writing of this book, however, they were being updated, among other things to take this pathogen into account. There are at present no procedural controls for prevention of HEV contamination of pork products, although on-farm controls including regular cleaning of pens, the use of easily cleanable barn materials, e.g. rubber/steel, and fly control, have been demonstrated to reduce transmission between pigs.

HACCP (Hazard Analysis Critical Control Points)

Hazard Analysis Critical Control Points (HACCP) is a widely established risk management concept within the food industry. It is based on the identification of hazards and a step-wise approach to controlling them (see Box 6.1). However, current HACCP systems have been formulated and

Box 6.1 The Seven Principles of HACCP

1) Identify the hazards associated with the food supply chain (e.g. microbial).
2) Identify the critical points at which the hazard can be reduced or eliminated (e.g. through disinfection).
3) Establish the critical limits above which the hazard should not be present.
4) Implement procedures to monitor the presence of the hazard (e.g. PCR-based detection methods).
5) Identify the measures that will be used to reduce or eliminate the hazard at the critical points (e.g. heat or chemical disinfection).
6) Establish procedures to verify that the measures used in step 5 are effective (e.g. by product testing).
7) Keep records of all measures taken during the HACCP process.

validated for foodborne bacterial hazards, and their applicability to the control of virus contamination requires confirmation. Furthermore, HACCP only applies after the primary production stage.

Procedural Controls for Prevention of Virus Contamination of Shellfish

Contamination of shellfish with viruses is extrinsic, with norovirus and HAV exposure occurring through contact with water contaminated with human faeces, mainly via sewage. Conventional sewage treatment is not fully effective for removing viruses, and treated sewage is usually discharged into environmental waters. It is known that a shellfish such as a mussel can filter more than 1000 l/day, and therefore if shellfish beds are located near to a sewage outflow, the shellfish which grow there have a high potential for virus contamination. Bivalve shellfish such as oysters are often eaten raw as a delicacy, and there have been numerous outbreaks of viral disease associated with this consumption practice. Consequently, the microbiological quality of the shellfish-growing water is a prime aspect to consider. Various national and international regulations exist which stipulate the parameters for the quality of the waters used for growing shellfish. These parameters are based on bacterial faecal indicators, such as coliforms, and do not relate to the presence of viruses. There are also regulations in some countries, e.g. the United States, regarding the discharge of human waste from shellfish-harvesting vessels. Compliance with the relevant regulations and careful selection of shellfish growing areas should be effective in reducing the risk of virus contamination.

Vaccination

Effective vaccines are available for HAV, and widespread vaccination of food handlers could be a very effective measure for reducing foodborne hepatitis A. However, this measure would be problematic to implement for several reasons, not least the expense involved, with the benefit possibly not justifying the cost. Vaccine development for preventing norovirus infection is challenging, especially given the lack of long-lasting immunity post-infection (see Chapter 2). There are currently no licensed vaccines available, but there are several that are under development, although their implementation could be equally as difficult as that for HAV. An HEV vaccine is currently only available in China and is not yet licenced internationally. An HEV vaccine as a control measure in pig production is unlikely.

Sources of Quoted Data

Effect of chlorine: Butot, S. et al. (2008). Effects of sanitation, freezing and frozen storage on enteric viruses in berries and herbs. *International Journal of Food Microbiology* 126: 30–35. Casteel, M.J. et al. (2008). Chlorine disinfection of produce to inactivate hepatitis A virus and coliphage MS_2. *International Journal of Food Microbiology* 125: 267–273. Gallandat, K. and Lantagne, D. (2020). A systematic review of chlorine-based surface disinfection efficacy against viruses. *American Journal of Infection Control* 48: 1116–1130. Girard, M. et al. (2010). Attachment of noroviruses to stainless steel and their inactivation, using household disinfectants. *Journal of Food Protection* 73: 400–404. Keswick, B.H. et al. (1985). Inactivation of Norwalk virus in drinking water by chlorine. *Applied and Environmental Microbiology* 50: 261–264. Mbithi, J.N. et al. (1990). Chemical disinfection of hepatitis A virus on environmental surfaces. *Applied and Environmental Microbiology* 56: 3601–3604.

Effect of ethanol: Escudero-Abarca, B.I. et al. (2022). Comparative assessment of the efficacy of commercial alcohol-based hand sanitizers against human norovirus. *Frontiers in Microbiology* 13: 869087. Kampf, G. et al. (2018). Efficacy of ethanol against viruses in hand disinfection. *Journal of Hospital Infection* 98: 331–342. Sato, S. et al. (2020). Alcohol abrogates human norovirus infectivity in a pH-adjusted system; some genotypes less affected. *Scientific Reports* 10: 72609.

Effect of plant materials: Su, X. and D'Souza, D.H. (2013). Grape seed extract for foodborne virus reduction on produce. *Food Microbiology* 34: 1–6. Randazzo, W. et al. (2017). Effect of green tea extract on enteric viruses and its application as natural sanitizer. *Food Microbiology* 66: 150–156.

Effect of heating: Cook, N. and van der Poel, W.H.M. (2015). Survival and elimination of hepatitis E virus: a review. *Food and Environmental Virology* 7: 189–194. Schilling-Loeffler, K. et al. (2025). Determination of hepatitis E virus inactivation during manufacturing of spreadable pork liver sausage and salami-like raw pork sausage. *International Journal of Food Microbiology* 429: 111018. Feagins, A.R. et al. (2008). Inactivation of infectious hepatitis E virus present in commercial pig livers sold in local grocery stores in the United States. *International Journal of Food Microbiology* 123: 32–37. Fuentes, C. et al. (2021). Inactivation of hepatitis A virus and human norovirus in clams by heat treatment. *Applied and Environmental Microbiology* 87: e01727–e01720. Robin, M. et al. (2019). Effect of natural ageing and heat treatments on GII.4 norovirus binding to histo-blood group antigens. *Scientific Reports* 9: 15312.

Effect of high hydrostatic pressure: Johne, R. et al. (2021). Stability of hepatitis E virus at high hydrostatic pressure processing. *International Journal of Food Microbiology* 339: 109013. Kingsley, D.H. et al. (2005). Pressure inactivation of hepatitis A virus in strawberry puree and sliced green onions. *Journal of Food Protection* 68: 1748–1751. Kingsley, D.H. et al. (2005). High-pressure inactivation of hepatitis A virus within oysters. *Applied and Environmental Microbiology* 71: 339–343. Nasheri, N. et al. (2020). Evaluation of high-pressure processing in inactivation of the hepatitis E virus. *Frontiers in Microbiology* 11: 461. Urtubia, A.S. et al. (2010). High-pressure inactivation of HAV within mussels. *Food and Environmental Virology* 2: 83–88; Wilkinson, N. et al. (2001). Resistance of poliovirus to inactivation by high hydrostatic pressures. *Innovative Food Science & Emerging Technologies* 2: 95–98.

Effect of UV light: Gómez-López, V.M. et al. (2021). Inactivation of foodborne viruses by UV light: a review. *Viruses* 13: 2229.

Depuration of shellfish: Stoppel, S.M. et al. (2023). Temperature dependent depuration of norovirus GII and Tulane virus from oysters (*Crassostrea gigas*). *Food and Environmental Virology* 15: 43–50.

Further Reading

Codex Alimentarius Commission (2012). Guidelines on the application of general principles of food hygiene to the control of viruses in food CAC/GL 79-2012.

Cook, N. et al. (2016). Persistence and elimination of human norovirus in food and on food contact surfaces: a critical review. *Journal of Food Protection* 79: 1273–1294.

Falcó, I. et al. (2024). Antiviral activity of natural compounds for food safety. *Food and Environmental Virology* 16: 280–296.

Govaris, A. and Pexara, A. (2021). Inactivation of foodborne viruses by high-pressure processing (HPP). *Foods* 10: 215.

Jama-Kmiecik, A. et al. (2021). Natural products and their potential anti-HAV activity. *Pathogens* 10: 1095.

Marriott, N.G. et al. (2018). *Principles of Food Sanitation*. Springer.

Sareen, S. (2016). A scheme and training manual on good agricultural practices. Food and Agriculture Organization (FAO). https://openknowledge.fao.org/handle/20.500.14283/i6677en

7

Future Challenges

Compared with other scientific disciplines, microbiology is young. Food microbiology is younger. Food virology is still in its infancy. Thus, whilst the discipline of food microbiology is reasonably well established, its virological area still requires considerable development.

One of the most challenging issues is the establishment of thoroughly effective methods for the detection of viruses in foods. Although some detection methods have been developed and standardised (see Chapter 4), further advances are needed to have fully fit-for-purpose procedures. The possible low level of virus contamination of food and the severe impracticality of their enrichment to increase virus numbers in a food sample, together with the somewhat optimistic expectation that methodologies should rapidly and unambiguously detect infectious particles, makes progress slow. Concurrently, fully developed systems for typing of viruses isolated from food to the strain level, and relating them to clinical isolates from outbreaks and cases, will provide essential information on the burden of foodborne viruses in public health. Currently, there are no globally harmonised schemes for typing enteric viruses, as there are for foodborne bacteria such as *Salmonella* or *Listeria*.

Currently, there is no agreement on bacterial or viral pathogens that could serve as indicator organisms of viral contamination of food, although several attempts utilising bacteriophages, *Escherichia coli*, or total coliforms have been made. A considerable research effort needs to be undertaken to identify suitable and effective viral indicators: these should not only indicate signal the potential presence of pathogenic viruses in foods, but also imply their infectivity. Meanwhile, a new approach has been proposed, aimed at the use of faecally derived human or animal viruses termed 'route-tracking viruses'. Although they are not considered as traditional foodborne viruses, their presence in foods or food supply chain environments would indicate that faecal contamination event(s) have occurred during food

Table 7.1 Route-tracking virus examples.

Route-tracking virus	Route tracked	Enteric viral pathogens with the potential for using route
Human adenovirus	From humans	HAV, NoV, astrovirus, poliovirus, rotavirus, etc.
Porcine adenovirus	From, e.g. pigs	HEV, porcine rotavirus
Bovine polyomavirus	From, e.g. cows	Bovine rotavirus

HAV, hepatitis A virus; HEV, hepatitis E virus; NoV, norovirus.

production, thus demonstrating that a route of contamination exists from the intestinal tracts of humans and/or animals to that point in the food supply chain where the route-tracking viruses are detected (Table 7.1).

As indicated in Chapter 5, a variety of such viruses may be necessary to fully demonstrate that the relevant route exists. Next-generation sequencing (NGS), with its ability to identify many microorganisms simultaneously, could have substantial potential for this, but considerable development of appropriate sample treatment procedures, to extract a highly concentrated and purified nucleic acid extract from a food sample and deliver it to the sequencing assay, will be necessary before this technology can be effectively deployed.

There is currently limited information on the prevalence of viruses in foods: what data we have now will only represent the tip of a very big iceberg. Harmonised activities for the establishment of surveillance systems of at-risk foods at national and international level need to be undertaken, in order to more fully determine the extent of virus presence in the relevant supply chains. This surveillance ideally should include not only monitoring at point of sale but, far more importantly, at primary production. It is unlikely that this will ever be done without formulation of regulations by the appropriate authorities, and for such regulation to be practical, the establishment of microbiological criteria for foodborne viruses will ultimately be necessary.

There are two types of food microbiological criteria – the Food Safety Criterion and the Process Hygiene Criterion. A food safety criterion defines the acceptability of a product or a batch of foodstuff placed on the market; in other words, the foodstuff does not present a hazard to the consumer through the presence of pathogens in numbers likely to cause infection. A process hygiene criterion defines the acceptable functioning of the production process; in other words, that the process is not allowing the ingress of pathogens. Implementation of these criteria will require effective monitoring systems to support their application. It is possible to speculate that, for example, a food safety criterion for norovirus in berries would be absence of the virus in say, 25 g, and that a process hygiene criterion for soft fruit production would be zero route-tracking viruses (or effective indicators) in samples of irrigation water and swabs of handlers' hands.

There is, unfortunately, substantial reluctance among governmental authorities and food business operators (FBOs) to establish regulations for foodborne viruses, even with the awareness that these pathogens are ranked as a high risk and have significant health and economic impacts, and the continuous production of data on the prevalence of viruses through food supply chains. Without regulations, there is no strong driver for FBOs to fully implement effective monitoring and control measures for foodborne viruses. This will only take place when regulations are in force. Without regulations, outbreaks will happen over and over again, and cases will continually occur. This situation can only be addressed by the relevant authorities and stakeholders, who will need to focus future research strategies towards setting of criteria and formulation of regulations.

Most of the traditional and emerging disinfection measures have been devised for use against bacterial contaminants, and their effectiveness for inactivation of viruses is uncertain. There is a growing body of experimental data on reduction of virus risks, but it was obtained without any harmonisation of the methodologies used to acquire it, and consequently very little of this knowledge has been used to implement virus-effective disinfection measures in the food industry. Harmonisation of inactivation studies should be the first step for achieving an effective goal of actual implementation in food supply chain settings. Studies should be performed to try to identify fully effective surrogates which will be affected by inactivation measures just the same as the pathogenic viruses are. Concurrently, considerable research is still required to have efficient cell culture systems which can be reproducibly established within the food analytical and research communities for use with foods rather than with only purified virus suspensions.

Special attention should be given to zoonotic viruses and their often-unidentified reservoirs. This is particularly relevant in the case of hepatitis E virus (HEV), for which the population exposure level is not known. It has been observed that rat HEV (*Rocahepevirus ratti*) can cause hepatitis E in humans. As rats are very common in farm environments, and rat HEV can be shed via their faeces, the possibility that this virus could infect or contaminate domestic pigs, and consequently be foodborne, is not inconceivable.

Population-level estimates of incidence, source attribution, and clinical impact of viruses are all needed to determine the burden of foodborne illness. In future, it may be possible that new foodborne pathogens may emerge or re-emerge, with new pathogen/food combinations. Their appearance should drive research towards recognition of the virus–host range, modes of virus transmission, environmental survival, and the role of food and water in disease spread. For example, snails are not frequently consumed, save in a few countries, therefore they could be recognised as a new food type. Snail meat may contain a number of pathogenic microorganisms; however, there is little data on the presence of enteric viruses in their tissues. Although human pathogenic viruses have not been detected in tissues of land snails, the presence of noroviruses and sapoviruses have been found in the digestive glands of marine snails. So far, only one outbreak has been reported, where consumption of raw marine snails (*Umbonium giganteum*) resulted in human norovirus infection. Also, there is no information on any role of crustaceans in transmission of enteric viruses to humans, although they have been detected in tiger and white shrimps.

There is currently much discussion about the environmental sustainability of meat production, and one suggested solution is to encourage the consumption of insect-based protein. An aspect of food insect production is that they are often cultivated on manure. Consequently, some risk of virus contamination of the final product could exist, e.g. if the manure is derived from animals which can be hosts of HEV. Some studies have demonstrated the presence of foodborne pathogenic bacteria such as salmonellae in insect products, and studies are required to investigate whether there is a similar potential for virus contamination.

As the incidence of foodborne viral disease became more apparent, it was recognised that the existing food safety standards were not as effective in managing the risk of foodborne viruses. This was broadly due to two factors: first, the different behaviour of viruses as foodborne hazards compared with bacteria and, second, a lack of data for modelling virus behaviour in foods due to challenges in detection and monitoring. In response, the Food and Agricultural Organisation (FAO) and the World Health Organisation (WHO) jointly convened an expert panel on the problem in 2007, with the Report of the panel's conclusions being published in 2008. Prominent among the conclusions of the panel was the need for more data to inform quantitative microbial risk assessments (QMRAs) for foodborne viruses. QMRA is a tool used in analysing the risk of foodborne infectious disease and supporting its management. There are four basic steps in a QMRA: Hazard Identification, Exposure Assessment, Dose Response, and Risk Characterisation. In simple terms, risk characterisation combines the information from the previous steps to define how many people will become ill from consumption of a pathogen-contaminated foodstuff. Hazard identification, which includes detailing the diseases caused by pathogens, is reasonably advanced for the majority of foodborne viruses. Dose response information is considerably less available, due to factors such as the impracticality of volunteer studies. And crucially, realistic exposure assessment data remains elusive as there have been insufficient monitoring and surveillance activities, especially using quantitative methods, to determine the prevalence of viruses in the food supply chain; and, as has been discussed in Chapter 4, the main methods used to detect viruses in foods do not provide information on infectivity. The outcomes of the few QMRAs which have been performed give at best only estimates of how many people are impacted by foodborne viral disease. More data, particularly exposure assessment, remain necessary for the development of fully informative virus QMRAs. This will require considerable further advancements in detection methods and developments in cell culturing for viruses like norovirus. Looking ahead, many emerging foodborne disease threats to come may be viruses, and viral QMRA will be an important tool for tackling them.

The emergence of unknown viruses with zoonotic potential could easily occur due to sustained close contact between human and animals. Such transmissions have been seen in the past, and they have also appeared recently in the case of SARS-CoV-2. Intensive animal farming or industrial livestock production could facilitate virus spread from one to another animal species. In addition, the mechanisms by which viruses evolve enable the formation of new strains, which in the presence of favourable conditions could quickly adapt to a new host. If novel strains with zoonotic potential are introduced to susceptible animal or human populations, their rapid spread may be manifested by the appearance of large numbers of infections, possibly with resulting fatalities. When they appear, the hazard needs to be quickly identified to prevent further disease spread. Therefore, not only should studies on the occurrence of currently known viral pathogens in livestock be conducted, but studies dealing with viral surveillance in wildlife are also essential to recognise novel viruses that may pose a threat to human and veterinary public health. It has been shown that transmission of emerging zoonotic viruses could occur not only through direct contact with infected animals, but also via the foodborne route during consumption of meals containing undercooked tissues of infected animals. It may be possible that new foodborne virus types could appear due to consumption of exotic meats originating from wildlife species. Monkeys are a common source of bushmeat, and although Ebola is not known to be foodborne, the handling and consumption of bushmeat have been associated with infection. Also, climate change may widen the natural geographical occurrence of previously restricted insect vectors of various diseases and allow them to carry pathogens to a wider population; for instance, there is growing evidence that tick-borne encephalitis has extended its geographical spread due to warming of the environment.

Climate change will also result in higher rainfall in certain areas, increasing the possibility of sewage overflows, resulting in the increased transfer of pathogenic microorganisms onto agriculture land. This will further result in an increased potential for crops such as leafy green vegetables to harbour foodborne viruses.

Wastewater surveillance can reveal the presence of several pathogenic virus types in the community. Not all are enteric: for example, analysis of wastewater samples for SARS-CoV-2 was routinely performed during and subsequent to the COVID-19 pandemic, to assist in determining the extent of circulation of this pathogen in specific regions. Recently, there have been reports of detection, at significant prevalence, of human papillomavirus (HPV) strains in wastewater. HPV can cause a variety of conditions of human epithelial tissues, some benign such as warts, and others more serious such as cancers. Transmission of these viruses is recognised as being mostly through person-to-person contact, and foodborne transmission is unknown and most probably highly unlikely. However, their presence in sewage, and their lack of an envelope and therefore potential environmental robustness, means that it is not totally unlikely that they could contaminate crops through flooding etc. This scenario also pertains for other viruses. Few if any studies are being performed to try to determine whether there is any significance of hitherto unrecognised viral threats in the food supply.

Food is essential for life, and therefore it should be free from disease-producing pathogens. Nevertheless, they can be occasionally, or frequently, found in food despite the efforts of the producers or manufacturers to prevent their presence. Generally, when food has been found to be contaminated, the contamination was unintentional and affecting only a limited volume of a foodstuff. A real threat will appear if the intentional release or contamination of food with biological agents takes place. Food could be targeted at several stages along the food supply chain. Agricultural bioterrorism could occur at primary production, when food crops or livestock are targeted. Food terrorism can also involve the contamination of food products during processing, manufacturing, storage, transport, distribution, or service. The US Centers for Disease Control and Prevention (CDC) have classified bioterrorism agents into several categories. Among viral agents present in the highest threat category, the most important are variola major (smallpox), Ebola and Marburg viruses, and arenaviruses (e.g. Lassa fever), which thankfully are not known to have a foodborne route of transmission. Nipah virus and tick-borne encephalitis virus (TBEV), however, do have such a route (see Chapter 2) and are classed as Category C bioterror agents. A potential bioterrorism or Nipah virus could furthermore be used as an agroterrorism agent, capable of causing significant economical loses in the agricultural industry: in an outbreak which occurred in Malaysia, the mortality rate in pigs was low, but to control the disease in pig farms and to prevent zoonotic virus transmission more than 1 million pigs from >900 farms were culled. Taking into account that poliovirus is highly contagious, easily released into food or water supplies, and can survive months in the environment, we should be aware of its use as a potential bioterror agent. Although the global vaccination program against poliovirus is nearly complete, outbreaks of polio still occur in certain parts of the world. Poliovirus is currently unlikely to be an attractive weapon for bioterror because much of the world has been vaccinated against it, and it is less lethal than viruses such as smallpox. However, if the international poliovirus vaccination strategy is concluded, then eventually the number of unvaccinated and immunologically naive people will increase rapidly making poliovirus a potential biological weapon. Deliberate contamination of foods or water sources by pathogenic viruses to produce mass casualties has thankfully not been attempted, but the ability to effectively monitor foodstuffs for viral contamination will be a useful resource in combating such food bioterrorism.

Capacity building is another challenge, particularly (though by no means exclusively) for low-income countries. As discussed in Chapter 4, the methods for analysis of foods for viruses require specialised facilities and trained and experienced analysts. In several countries, including some high-income countries, facilities for rapid deployment of foodborne virus detection methods are scarce or lacking, and this will have implications for management of large-scale outbreaks and also impact international trade in food commodities.

Food virology is a young discipline, although it has undergone substantial development since its inception. Nonetheless, its challenges remain considerable. To ensure that the threat to public health posed by foodborne viruses is efficiently combated, these challenges must be addressed as the discipline progresses towards maturity. It is hoped that through the brief introduction to the field which has been provided in this book, the reader will have gained a thorough insight into the issues surrounding virus contamination of the food supply chain and a working understanding of the measures by which these issues may in time be addressed.

Sources of Quoted Data

Norovirus infection caused by consumption of marine snails: Tanaka, T. et al. (2015). Detection and genetic analysis of noroviruses and sapoviruses in sea snail *Umbonium giganteum*. *Applied and Environmental Microbiology* 81: 6957–6965.

Human papillomavirus in wastewater: Mejias-Molina, C. (2024). Wastewater-based epidemiology applied at the building-level reveals distinct virome profiles based on the age of the contributing individuals. *Human Genomics* 18: 10.

Nipah virus outbreak in Malaysia: Mohd Nor, M.N. et al. (2000). Nipah virus infection of pigs in peninsular Malaysia. *Revue Scientifique et Technique (International Office of Epizootics)* 19: 160–165.

Further Reading

Brul, S. et al. (2012). 'Omics' technologies in quantitative microbial risk assessment. *Trends in Food Science & Technology* 27: 12–24.

Food and Agricultural Organisation (FAO) & World Health Organisation (WHO) (2021). Microbiological risk assessment – guidance for food. Microbiological Risk Assessment Series No. 36. Rome.

Green, M.S. et al. (2018). Confronting the threat of bioterrorism: realities, challenges, and defensive strategies. *Lancet Infectious Diseases* 16: e2–e13.

Howley, P.M. (2020). *Fields Virology: Emerging Viruses*. Lippincott Williams & Wilkins.

International Commission on Microbiological Specifications for Foods (2011). *Microorganisms in Foods 8: Use of Data for Assessing Process Control and Product Acceptance*. Springer.

Rasco, B.A. and Bledsoe, G.B. (2019). *Bioterrorism and Food Safety*. CRC Press.

Scaffardi, L. and Formici, G. (eds.) (2022). *Novel Foods and Edible Insects in the European Union: An Interdisciplinary Analysis*. Springer.

Index

Printed and bound by CPI Group (UK) Ltd, Croydon, CR0 4YY
17/09/2026
14964294-0001